DISCARD

Conversations with Carl Sagan

Literary Conversations Series
Peggy Whitman Prenshaw
General Editor

Photo credit: © Photofest

Conversations with Carl Sagan

Edited by
Tom Head

University Press of Mississippi
Jackson

www.upress.state.ms.us

The University Press of Mississippi is a member of the Association of American University Presses.

Copyright © 2006 by University Press of Mississippi
All rights reserved
Manufactured in the United States of America

First edition 2006
∞
Library of Congress Cataloging-in-Publication Data

Sagan Carl, 1934–
    Conversations with Carl Sagan / edited by Tom Head.— 1st ed.
       p. cm. — (Literary conversations series)
    Includes bibliographical references and index.
    ISBN 1-57806-735-9 (cloth : alk. paper) — ISBN 1-57806-736-7 (paper : alk. paper)
    1. Sagan, Carl, 1934– —Interviews. 2. Novelists, American—20th century—Interviews. 3. Science fiction—Authorship. I. Head, Tom. II. Title. III. Series.

PS3569.A287Z475   2006
520′.92—dc22
[B]   2005048747

British Library Cataloging-in-Publication Data available

Books by Carl Sagan

The Atmospheres of Mars and Venus. Washington, DC: National Academy of Sciences, 1961. With W[illiam] W. Kellogg.
Intelligent Life in the Universe. Oakland, CA: Holden-Day, 1963. With I[osif] S[amuilovich] Shklovskii.
Planets. New York: Time-Life Science Library, 1966. With Jonathan Leonard.
Planetary Exploration: The Condon Lectures. Corvallis, OR: University of Oregon Press, 1970.
Planetary Atmospheres. New York: Reidel, 1971. Edited with Tobias C. Owen and Harlan J. Smith.
Space Research XI. 2 vols. Berlin: Akademie Verlag, 1971. Edited with K[irill] Ya. Kondratyev and M[ichael] Rycroft.
UFOs: A Scientific Debate. Ithaca, NY: Cornell University Press, 1972. Edited with Thornton Page.
Communication with Extraterrestrial Intelligence, editor. Cambridge, MA: MIT Press, 1973.
Mars and the Mind of Man. New York: Harper, 1973. With Ray Bradbury, Arthur C. Clarke, Bruce Murray, and Walter Sullivan.
Life Beyond Earth and the Mind of Man. Washington, DC: U.S. Government Printing Office, 1973. With Richard Berendzen, Ashley Montagu, Philip Morrison, Krister Stendhal, and George Wald.
The Cosmic Connection: An Extraterrestrial Perspective. New York: Doubleday, 1973.
Other Worlds. New York: Bantam, 1975.
The Dragons of Eden: Speculations on the Evolution of Human Intelligence. New York: Random House, 1977.
Murmurs of Earth: The Voyager Interstellar Record. New York: Random House, 1978. With F[rank] D[onald] Drake, Ann Druyan, Jon Lomberg, Linda Sagan, and Timothy Ferris.
Broca's Brain: Reflections on the Romance of Science. New York: Random House, 1979.
Cosmos. New York: Random House, 1980.
The Cold and the Dark: The World After Nuclear War. New York: Norton, 1984. With Paul R. Ehrlich, Donald Kennedy, and Walter Orr Roberts.
Comet. New York: Random House, 1985. And Ann Druyan.

Contact: A Novel. New York: Random House, 1985.
A Path Where No Man Thought: Nuclear Winter and the End of the Arms Race. New York: Random House, 1989. With Richard Turco.
Shadows of Forgotten Ancestors: A Search for Who We Are. New York: Random House, 1992. And Ann Druyan.
Pale Blue Dot: A Vision of the Human Future in Space. New York: Random House, 1994.
The Demon-Haunted World: Science as a Candle in the Dark. New York: Random House, 1995.
Billions and Billions: Thoughts on Life and Death at the Brink of the Millennium. New York: Random House, 1997.

Contents

Introduction ix
Chronology xxi
A Conversation with Carl Sagan *Timothy Ferris* 3
A Resonance with Something Alive *Henry S. F. Cooper, Jr.* 19
Carl Sagan Interviewed *Joseph Goodavage* 36
Second View: Sagan on *Encounters* *Art Harris* 47
Carl Sagan's *Cosmic Connection* and Extraterrestrial Life-Wish *Dennis Meredith* 51
The Cosmos *Jonathan Cott* 57
God and Carl Sagan: Is the Cosmos Big Enough for Both of Them? *Edward Wakin* 68
A Pale Blue Dot *Claire Marino* 76
Sagan: Dump Environmentally Unconscious Slobs *Ponchitta Pierce* 79
Talk of the Nation: Science Friday *Ira Flatow* 82
Bringing Science Down to Earth *Anne Kalosh* 99
The Charlie Rose Show *Charlie Rose* 106
A Slayer of Demons *Psychology Today* 113
Talk of the Nation: Science Friday *Ira Flatow* 125
The Charlie Rose Show *Charlie Rose* 141
The Final Frontier? *Joel Achenbach* 151
Index 163

Introduction

"When you're in love," Carl Sagan said in his last published interview,[1] "you want to tell the world. I've been in love with science, so it seems the most natural thing in the world to tell people about it." The passion for science Sagan displayed in his interviews was not based strictly on professional interest; it was based on a deeply personal and sincere curiosity about the cosmos and an almost religious commitment to preserving the vulnerable humanity he saw within it. Before Sagan, scientists had argued for centuries that the Earth is not at the center of the universe and existentialists had argued for centuries that human life is not subject to special protection, but these were abstract ideas generally described in an abstract way. These beliefs, though widely accepted by philosophers and scientists alike, had not been introduced together into the public consciousness as a clear, meaningful model of humanity and the cosmos that could be visualized and understood. Sagan attempted to combine these images and emerged with an original way of describing the universe. As he told *Interview* magazine in 1996:

> We live on an obscure hunk of rock and metal circling a humdrum sun, which is on the outskirts of a perfectly ordinary galaxy comprised of 400 billion other suns, which, in turn, is one of some hundred billion galaxies that make up the universe, which, current thinking suggests, is one of a huge number—perhaps an infinite number—of other closed-off universes. From that perspective, the idea that we're at the center, that we have some cosmic importance, is ludicrous.[2]

Sagan's interviews convey a humanism, a confidence in the potential of the human race tempered by a sense of its vulnerability as a tiny hive swept about in an indifferent cosmos. This explains, in part, his passion for science—the only way humanity has ever learned to protect itself from the unspeakably powerful and mysterious blind forces of nature. "It's not that scientists are prejudiced toward science," Sagan told *U.S. News*' Stephen

Budiansky. "It's that science demonstrably works better than anything else. If something else worked better, we'd be for the something else."[3]

But if Sagan described great danger in a frighteningly large and mysterious universe, he also found wonder in it. He seemed to convey a natural tolerance for doubt and ambiguity, and stood in awe of a universe that holds lifetimes of mystery within it and might one day be understood. He told *Interview* that "there's something terribly beautiful, austere, glorious, majestic about the fact that the same laws apply everywhere."[4] Sagan's sense of vulnerability and doubt was always held in tension with the wonder he felt at the majesty of the universe and the excitement he felt with each new astronomical discovery. "If you look into science," he told Edward Wakin in a 1981 interview, "you will find a sense of intricacy, depth, and exquisite beauty which, I believe, is more powerful than the offerings of any bureaucratic religion."

Carl Sagan felt that this sense of curiosity and wonder about the universe was natural in children. "Every kid starts out as a natural-born scientist," he told *Psychology Today* in 1996, "and then we beat it out of them." When five-year-old Carl visited the 1939 New York World's Fair, he was deeply impressed by its stories of an amazing universe and an inspiring human future within it. His interest only grew as he got older. Sagan told *Highlights for Children* that "with an early bedtime in the winter, I could look out my window and see the stars, and the stars were not like anything else in my neighborhood."[5] When he was nine years old, he wanted to learn more about these stars and asked his mother to explain them to him. She suggested that he use his new library card to do some research on his own, and he soon found himself sitting down with a book on astronomy. This afternoon spent reading a book, an event forgettable in the lives of most children, triggered something in Sagan that would define his interests for the rest of his life. "It was in that library," he told *Highlights*, "reading that book, that the scale of the universe opened up to me. There was something beautiful about it."[6]

Immersed in real science, comic books, and the Mars novels of Edgar Rice Burroughs, Sagan envisioned an exciting universe full of intelligent life and unfathomable mysteries. It was during this very early phase of his life that the passion underlying the scientific position he would become best known for began to take its shape. Some in Sagan's position would have attempted to clearly divide the adult scientific interest in extraterrestrial life from childhood fantasies, but Sagan had no difficulty suggesting that his adult passions and childhood passions were in many respects very similar. Sagan's devotion to science was rooted in a childlike wonder, and he saw nothing unsophisticated about this sense of wonder. "There is a danger of underestimating the intelligence of kids," he told *New York Times* reporter Boyce Rensberger in 1977. "Kids can understand some pretty deep things."[7]

Decades later, in June 1960, twenty-six-year-old Carl Sagan received his Ph.D. in astronomy and astrophysics from the University of Chicago. His dissertation, titled *Physical Studies of the Planets*, speculated on the possibility of extraterrestrial life and the conditions that could allow it to survive. Though married and raising a son of his own, Carl Sagan the educated scientist shared many of the same interests as Carl Sagan the curious child. At the same time, years of education in astronomy and the scientific method had given him a heavy dose of skepticism. For most of his public life, it was that tension between wonder and skepticism that made him a successful advocate for science. In order to appeal to nonscientists, he had to convey a sense of wonder; in order to be faithful to the science he described, he had to be a skeptic. In this respect he was a human reflection of the problem science poses. According to Sagan, effective science must be tentative, but only to a point—there comes a time when it is necessary to commit to an idea. "What we need for survival," he told Wakin, "is a well-tuned mix of creativity and doubt."

Although he was enthusiastic about the possibility of extraterrestrial life, Sagan found the hypothesis of alien visitation to be highly unlikely. Evidence supporting the idea that UFOs are alien spacecraft is, Sagan told *Interview*, "wimpy—I mean, nothing close to what would convince a court of law, much less science."[8] He found alien abduction accounts (involving purported extraterrestrials who, as he put it in a 1994 interview with Anne Kalosh, always seemed to be "short, sullen, grumpy, and sexually obsessed") to be even more dubious. By the early 1970s, Sagan had long established both the speculative and skeptical aspects of his persona.

During this time, Sagan slowly and almost accidentally developed into a public figure. He was outspoken, charismatic, direct, and assertive in a way that scientists generally were not, spoke freely about an interesting area of science (exobiology), and did not shy away from public attention. When *Time* magazine published a feature on the possibility of life on Mars, it quoted Sagan throughout simply because he produced more interesting quotes than any other scientist they profiled. When his exobiology-themed book *The Cosmic Connection* was published in 1973, he was ready for prime time—or, at least, late night, showing up on Johnny Carson's *The Tonight Show* for multiple interviews and specials (including a passionate and engaging thirty-minute crash course on astronomy). This resulted in over a half million sales for the book, and turned Sagan into America's best-known scientist within a matter of weeks. He followed up his success with a lengthy interview in *Rolling Stone*, which attracted readers who were not generally interested in lengthy interviews with astronomers. Sagan had made science more popular and, in the process, established the public role that would define him for the rest of his life. As Sagan biographer William Poundstone writes:

> This was a natural outgrowth of his earlier attempts to reconcile science and the youth culture. At a time when science was suspect among much of the population, this coverage forged a reputation for Sagan as a socially conscious, iconoclastic scientist. . . . Scientists were not normally featured in *Rolling Stone*, but [interviewer Timothy] Ferris sensed, correctly, that Sagan would make a good interviewee.[9]

When NASA designed the deep space *Voyager* probes for launch in 1977, Sagan was asked to chair a committee to define the contents of the Golden Disc, a sort of time capsule for the benefit of any extraterrestrials who might run across the craft years, centuries, or millennia later. The disc displayed the position of Earth's sun within the Milky Way galaxy (relative to the location of easily-identifiable pulsars) and featured both audio and video content: ninety minutes of music (ranging from Bach to Japanese flute music) and about one hundred full-color photographs covering many aspects of Earth's history and geography, biology and the evolutionary process, and human life.

Yet the project that Sagan is best remembered for is *Cosmos*, a fourteen-hour miniseries aired by PBS in 1980. For two years Sagan constantly worked on the miniseries and its accompanying book; although he had a large staff and an $8.2 million budget, the research and filming involved in the project left Sagan with very little free time. His personal life during this period was also particularly stressful; he and his second wife were negotiating a divorce, and his father was dying of lung cancer.

In a rare 1979 interview, conducted by Dennis Meredith for *Science Digest*, Sagan outlined his goals for the series: "I'd like the series to be so visually stimulating that somebody who isn't even interested in the concepts will watch just for the effects. And I'd like people who are prepared to do some thinking to be *really* stimulated." When *Cosmos* premiered in September 1980, it quickly became the highest-rated PBS series ever and the accompanying book spent seventy weeks on the *New York Times* best-sellers list. Sagan's turtleneck and tan corduroy jacket, his mellifluous voice recorded over existential chords, and the sweeping camera shots of surreal astronomical landscapes defined Sagan for the rest of his life.

Johnny Carson parodied the series with a plosive "billions and billions" (a phrase that never appeared in *Cosmos*) and, while the series was generally popular, some critics felt that Sagan seemed entirely too arrogant. For his part, Sagan blamed this largely on odd visuals. The miniseries used a metaphorical device, a giant dandelion seed called the "Spaceship of the Imagination." Sagan considered this device somewhat hokey, and believed that it—combined with the reaction shots he described in a 1985 *New York Times* interview as "interminable close-ups of me looking awed"[10]—was largely responsible for creating the impression that he was in love with his own intelligence. Sagan was not thrilled with these aspects of the production, but the final decision did not rest with him; the film was in danger of running over budget, and reaction shots could be used to fill gaps far more cheaply than the additional special effects the production team had expected to use.

The arrogance charge was not limited to *Cosmos*, however. Sagan gave no deference to the metaphysical doctrines proposed by organized religion in his books or in his miniseries, and at times he attacked them outright. In a 1981 interview with *U.S. Catholic* conducted by Edward Wakin and titled "God and Carl Sagan: Is the Cosmos Big Enough for Both of Them?," Sagan tried to explain his position in a conciliatory way without surrendering any ground:

> I think religion has something to say to science about the social underpinnings of the enterprise of science, something about the goals of science, the human values that should always be in mind when we do science. . . . I also think science has a fair amount to say to religion, mainly about the nature of evidence. . . . I am concerned that the authoritarian aspect of religion poses real dangers for our survival.

For his part, Sagan did not belong to any organized religious tradition. "To be certain of the existence of God and to be certain of the nonexistence of God," he told *U.S. Catholic*, "seem to me to be the confident extremes in a subject so riddled with doubt and uncertainty as to inspire very little confidence indeed." When asked in a 1996 interview what his religious beliefs were, Sagan gave a direct answer: "I'm agnostic."[11] He made little of his agnosticism, however, because he felt that terms referring to theism, agnosticism, and atheism were nearly meaningless. As he told National Public Radio's Terry Gross, host of *Fresh Air*: "I find that you learn absolutely nothing about someone's belief if you ask them 'Do you believe in God?' and they say yes or no. You have to specify which of the countless kinds of God you have in mind."[12] He also took issue with the stereotype of the arrogant scientist attacking the humble religious leader; "science," he told Wakin, "is humble. It doesn't impose its own views on the universe." When Sagan spoke to longtime friend Linda Obst in a February 1996 interview, he remarked that there seems to be an uncanny elegance to the nature of the universe and expressed some sympathy towards those who believe in a traditional God:

> [Elegance] goes directly to the question of how the laws of nature are constructed. Nobody knows the answer to that. Nobody! It's a perfectly legitimate hypothesis, in my view, to say that some extremely elegant creator made those laws. But I think if you go down that road, you must have the courage to ask the next question, which is: Where did that creator come from? And where did his, her, or its elegance come from? And if you say it was always there, then why not say that the laws of nature were always there and save a step?[13]

Sagan dedicated himself to social activism during his later career, campaigning for environmentalism and nuclear disarmament. Here he relied on his earliest work as an astronomer, where he argued the hypothesis—radical then, but now widely accepted—that Venus's hot surface temperature can be attributed to the greenhouse effect. "When uninformed, politically motivated radio talk-show hosts say that the greenhouse effect is a hoax," he told *Interview*, "we should point them toward Venus to see what a real greenhouse effect is like—it's a very good reality check."[14] Sagan found it a potent image and was frustrated by political apathy on environmental issues. "It's much easier to demonize the head of a foreign nation, especially one from a different culture than ours," he told interviewer Ponchitta Pierce at 1992's United Nations Earth Summit. "It is much more difficult to raise public concern about invisible gases." His arguments in favor of nuclear disarmament were in some respects more influential; he was one of five scientists who popularized the phrase and concept of nuclear winter in a 1983 article for *Science* magazine, and the terrifying idea of a world made inhospitable by massive clouds of ash had a great deal of resonance. Both of these causes were incorporated into Sagan's larger philosophy of reconciliation and world peace, grounded—as so many of his ideas were—in the vulnerability of humanity.

Other causes piqued Sagan's interest as well. Throughout the 1980s and 1990s, Sagan advocated increased funding for science, particularly astronomy. "We send people 200 miles up in a tin can, report that the newts are reproducing nicely, thank you, and then we're told this is NASA at the forefront of exploration. It's more than 200 miles between New York and Boston," Sagan complained in a 1996 interview. "I mean, let's explore."[15] He was also concerned about the state of science education. "There are just too many cases where the science teacher merely hands down, as if from Mount Sinai, the findings of science without giving any idea of the method by which that information was acquired."[16] This sort of situation can create problems in any nation but especially, Sagan felt, in an industrialized democracy. "We live in a society absolutely dependent on science and technology," he told Anne Kalosh in 1995, "and yet have cleverly arranged things so that almost no one understands science and technology. That's a clear prescription for disaster."[17] Science literacy is useful not only for understanding science and technology itself, Sagan argued, but also for developing the sort of critical thinking skills that are useful in an open society: "Democracy and science share skeptical attitudes," he told *U.S. News and World Report*, "a sense that arguments from authority are bankrupt, that you must demonstrate to skeptics the validity of the point you're making."[18]

Sagan summed up both of these concerns in his book *Pale Blue Dot: A Vision of the Human Future in Space* (Random House, 1994), titled after a 1990 *Voyager* photo taken from the edge of the solar system that showed Earth to be a tiny, almost invisible blue speck. The book was a summary of Sagan's global ethic and moral vision—everything he had ever said about the destiny of humanity was represented in this book, which argued that Earth is extremely vulnerable, that everything meaningful to us can be annihilated at any time by an unthinking humanity or an unfeeling cosmos. Referring to the *Voyager* photograph, he wrote:

> Look again at that dot. That's here. That's home. That's us. On it everyone you love, everyone you know, everyone you ever heard of, every human being who ever was, lived out their lives. . . . Think of the rivers of blood spilled by all those generals and emperors so that, in glory and triumph, they could become the momentary masters of a fraction of a dot. . . . Our planet is a lonely speck in the great enveloping cosmic dark. In our obscurity, in all this vastness, there is no hint that help will come from elsewhere to save us from ourselves. . . . It has been said that astronomy is a humbling and character building experience. There is perhaps no better demonstration of the folly of human conceits than this distant image of our tiny world.[19]

With the threats posed by nuclear, chemical, and biological weapons, and environmental damage, Sagan saw little long-term hope for Earth unless the human race learns to work together in peaceful cooperation. "Some might not like the idea of a global solution," he conceded in a 1985 interview, "but there's no way out. Our technology has guaranteed that only global solutions are going to work."[20] Likewise, he also saw the threat posed by a sometimes hostile universe; it is almost certain that a meteor of the size that struck Jupiter during the Shoemaker-Levy in March 1995 would wipe out most life on Earth and potentially end the human race. Sagan believed that these threats could be addressed through a mix of international cooperation, nuclear disarmament, environmental progress, and substantial increases in science funding. Although Sagan believed that the challenges facing the human race are dire, he argued that humanity as a whole is making progress and may stand a good chance of long-term survival if it continues to do so. As he told *U.S. News and World Report* in 1985:

> [Y]ou can see a very clear long-term trend if you just look at the size of the group the average person identifies with. One hundred thousand years ago people identified with groups of hunter-gatherers—maybe 100 people. Today the typical allegiance is to tens or even hundreds of millions of people.[21]

While Sagan argued for the future survival of humanity, his own survival faced a new threat. In December 1994, he was diagnosed with myelodysplasia, a rare bone marrow disease resulting from faulty, rapidly spreading stem cells that produce inadequate numbers of red blood cells. He received three bone marrow transplants from his sister in April 1995, and for over a year he appeared to be in remission.

Sagan spent much of this time focusing on his work. He promoted the global ethic he had proposed in *Pale Blue Dot*, attending many conferences, interviews, and symposiums when his health permitted it, co-wrote several new scientific papers, assisted with the production of an upcoming film based on his 1985 novel *Contact*, signed up to resume his teaching schedule at Cornell University, and finished up a new book called *The Demon-Haunted World: Science as a Candle in the Dark* (Random House, 1995). In *The Demon-Haunted World*, Sagan wrote a manifesto on the scientific method, providing a step-by-step hypothetical reasoning primer called "the baloney detection kit" that readers could use to evaluate questionable claims. In the book, Sagan resurrected the Enlightenment metaphor of reason as a candle shining into the darkness of the universe, empowering individual human beings to think freely and take control of their own destinies.

Pseudoscientific concepts such as astrology, crystal healing, and alien abduction were, in Sagan's view, ultimately mind-numbing appeals to authority. Sagan argued that by countering fuzzy thinking with airtight reasoning, human beings can foster a scientific attitude and thereby increase the odds that humanity will be able to accept new ideas and solve the considerable problems it faces. Although some of his critics argue that the scientific method also restricts new ideas, Sagan scoffed at the notion. "I don't think that scientists are prejudiced to begin with," he told *PBS Nova*. "Prejudice means pre-judging. They're post-judiced. After examining the evidence, they decide there's nothing to it."[22]

The film version of *Contact* was in production, starring Jodie Foster in the role of scientist Ellie Arroway. The film, like the novel, tells a story about Earth's first contact with intelligent extraterrestrials and the impact that contact might have on humanity. "[I]f we got a message [from extraterrestrials] it would have to be from somebody much smarter than us, because anybody dumber than us is too dumb to send a message—we've just invented radio," Sagan told Ira Flatow in 1994. "That means that every branch of human knowledge is now up for reconsideration. . . . Did we get something wrong in fundamental astronomy? Did we make a mistake in mathematics somewhere? You can see people being really nervous, but the chance to tap into such knowledge—it's like going to school for the first time." Upon its release in 1997, *Contact* was cheered by critics and earned $171.1 million worldwide—but Sagan did not live to see its release.

By December 1996, the myelodysplasia and the bone marrow transplants had taken their toll. Even on his deathbed, Sagan was rigorously honest about human vulnerability. His vital signs had boosted when his children came to visit him in his hospital room, but he knew his chances of survival. "This is a deathwatch," he privately told his wife and collaborator Ann Druyan. "I'm going to die." "No," she told him. "You're going to beat this, just as you have before when it looked hopeless." "Well," he responded, "we'll see who's right about this one."[23] He died from pneumonia near midnight on December 20, 1996.

Sagan's relatively early death at the age of sixty-two stunned his fans and colleagues. He had not even retired from teaching, and numerous posthumous research articles followed shortly after his death. Yet the same sense of perspective that informed his vision of a vulnerable humanity also led him to feel lucky that he had lived a life where he had the opportunity to accomplish what he had accomplished. Even as early as 1973, he described himself as fortunate for having been able to work on the Mars *Mariner* and the other planetary exploration projects. As he wrote in *The Cosmic Connection*:

> Had I been born fifty years earlier, I could have pursued none of these activities. They were all then figments of the speculative imagination. Had I been born fifty years later I also could not have been involved in these efforts, except possibly the [search for extraterrestrial life]. . . . I think myself extraordinarily fortunate to be alive at the one moment in the history of mankind when such ventures are being undertaken.[24]

Sagan's life came with several curious postscripts, and perhaps the one most relevant to his public life was the book he was working on at the time of his death. For almost twenty years, he had been associated with Johnny Carson's comedic phrase "billions and billions," even though Sagan himself had never used it, and quickly grew very irritated with it. Sagan's working title for the new book, published in 1997, was *Billions and Billions*. At the end of his life, Sagan had come to terms with both the serious and less-than-serious aspects of his public persona.

The first interview Sagan ever gave was an unpublished 1966 six-hour taped conversation that is currently housed at the American Institute of Physics. In the years since, many of Sagan's interviews have become iconic; the *Rolling Stone* interview in 1973, for example, is frequently cited by biographers as playing a major role in transforming Sagan into a public figure. The sixteen interviews collected in this book span a twenty-six-year period, from 1973 to 1996, and are organized chronologically. As with all books in the Literary Conversations series, the interviews are reproduced as they originally appeared and have not been edited in any significant way. Most deal with Sagan's most recent project at the time of the interview, but some focus on more specific themes such as environmentalism, religion, and the possibility of extraterrestrial life.

This book is, in every meaningful sense of the word, a collaboration. It is a collection of Carl Sagan's writings, not mine, and my greatest hope is that this selection of interviews does justice to the life he lived, the work he produced, and the human values he affirmed. It is necessary to approach any book of this kind as an objective scholar, but I approached this volume as both a researcher and a fan; Sagan's public career was so vast that much of his work profoundly influenced my own intellectual development. I am in debt to my editors, Anne Stascavage and Seetha Srinivasan at the University Press of Mississippi, whose sensible advice and gentle encouragement guided the production of this book at every stage. I also owe thanks to assistant editor Walter Biggins at UPM, Laurie Harper at Sebastian Literary Agency, Charlie Brenner at the Jackson/Hinds Library System, and Shane Hunt at 4ResearchSolutions.com for their invaluable assistance in preparing this manuscript. This book would not have been possible without the generous permission granted by those who own copyright to the interviews reprinted here. As always, I would also like to extend special thanks to my family for their love and support.

TH

Notes

1. Jack Rightmyer, "Stars in His Eyes," *Highlights for Children*, January 1997.
2. Linda Obst, "Valentine to Science," *Interview*, February 1996.
3. Stephen Budiansky, "Keeper of the Flame," *U.S. News and World Report*, March 18, 1996.
4. Obst.
5. Rightmyer.
6. Ibid.

7. Boyce Rensberger, "Carl Sagan: Obliged to Explain," *New York Times*, May 29, 1977.
8. Obst.
9. William Poundstone, *Carl Sagan: A Life in the Cosmos*. New York: Holt, 1999.
10. Glenn Collins, "The Sagans: Fiction and Fact and Back Again," *New York Times*, September 30, 1985.
11. Jim Dawson, "The Demon-haunted World," *Minneapolis Star-Tribune*, March 2, 1996.
12. Terry Gross, *NPR Fresh Air*, May 29, 1996.
13. Obst.
14. Ibid.
15. Budiansky.
16. Ibid.
17. Anne Kalosh, "An Interview with Carl Sagan," *The Planetarian*, March 1995.
18. Budiansky.
19. Carl Sagan, *Pale Blue Dot*, New York: Random House, 1994, pp. 8–9. Quoted by permission of the Estate of Carl Sagan.
20. "Today's Technology May Find E.T. If He's Out There," *U.S. News and World Report*, October 21, 1985.
21. Ibid.
22. "Kidnapped by UFOs?: Interview with Carl Sagan," PBS NOVA Online, www.pbs.org/wgbh/nova/aliens/carlsagan.html
23. Ann Druyan, epilogue to *Billions and Billions: Thoughts on Life and Death at the Brink of the Millenium* by Carl Sagan. New York: Random House, 1997. Quoted by permission of the Estate of Carl Sagan.
24. Carl Sagan, *The Cosmic Connection*, New York: Doubleday, 1973, p. viii. Quoted by permission of the Estate of Carl Sagan.

Chronology

1934	9 November: Carl Edward Sagan is born to Samuel Sagan and Rachel Gruber Sagan in New York, New York.
1954	Awarded a Bachelor of Arts, with honors, from the University of Chicago.
1955	Awarded a second bachelor's degree from the University of Chicago (a Bachelor of Science in physics), and begins graduate school.
1956	Awarded a Master of Science in physics from the University of Chicago.
1957	Marries Lynn Alexander shortly after she graduates from the University of Chicago. Sagan's article "Radiation and the Origin of the Gene," printed in the journal *Evolution*, is his first publication.
1959	Son, Dorion, is born.
1960	Awarded a Ph.D. in astronomy and astrophysics from the University of Chicago. His dissertation, *Physical Studies of the Planets*, deals in some depth with the possibility of extraterrestrial life. Son, Jeremy, is born.
1960–1962	Serves as a Miller Research Fellow in astronomy at the University of California, Berkeley, where his wife, Lynn, is working on a Ph.D. in biology.
1961	*The Atmospheres of Mars and Venus: A Report by the Ad Hoc Panel on Planetary Atmospheres of the Space Science Board* (coauthored with W. W. Kellogg) is published by the National Academy of Sciences. By this time Sagan has already become known for championing the theory that Venus's hot atmosphere came about by way of the greenhouse effect, a view that would later play a significant role in his environmental activism.

1962–1968	Serves as assistant professor of astronomy at Harvard University, as a resident astrophysicist at the Smithsonian Institution, and as associate editor of *Icarus: The International Journal of Solar System Studies*.
1963	*Intelligent Life in the Universe* (coauthored with Soviet scientist I. S. Shklovskii) is published. Sagan and Lynn Alexander are divorced, and Alexander finishes her Ph.D. in biology from the University of California at Berkeley the same year. As Lynn Margulis, she will become a well-known biologist, author of thirty-one books and more than eighty peer-reviewed articles.
1966	Time-Life Science Library volume *The Planets* (coauthored by freelance writer Jonathan Leonard) is published.
1966	February: Serves as a member of the Ad Hoc Committee to review the U.S. Air Force's Project Blue Book, a documented history of UFO sightings. The committee concludes that "in 19 years and more than 10,000 sightings recorded and classified, there appears to be no verified and fully satisfactory evidence of any case that is clearly outside the framework of presently known science and technology."
1968	Marries Linda Salzman, an artist.
1968–1970	After leaving Harvard, serves as an associate professor at Cornell University.
1968–1979	Promoted to editor-in-chief of *Icarus: The International Journal of Solar System Studies*.
1970	*Planetary Exploration: The Condon Lectures* is published from a series of lectures Sagan delivered on astrogeology at Oregon State University. Receives NASA's Apollo Achievement Award. Son, Nicholas, is born.
1970–1976	Full professor of astronomy and space sciences at Cornell University.
1971	Publication of *Planetary Atmospheres* (an anthology co-edited with Tobias Owen and Harlan J. Smith) and *Space Research XI* (a two-volume anthology co-edited with Kirill Kondratyev and Michael Rycroft).
1972	Publication of *UFOs: A Scientific Debate* (an anthology co-edited with fellow astronomer Thornton Page).

1973	Publication of *Communication with Extraterrestrial Intelligence* (an anthology edited by Sagan), *Mars and the Mind of Man* (with Ray Bradbury, Arthur C. Clarke, Bruce Murray, and Walter Sullivan), *Life Beyond Earth and the Mind of Man* (with Richard Berendzen, Ashley Montagu, Philip Morrison, Krister Stendhal, and George Wald), and *The Cosmic Connection: An Extraterrestrial Perspective*. Although Sagan is already a prolific writer and something of a celebrity scientist, it is not until *The Cosmic Connection*, a breezy and eloquent defense of the value of space exploration in general and SETI in particular, that he begins to become the popular ambassador to science that he would remain for the rest of his career.
1975	Serves as astronomy section chair for the American Association for the Advancement of Science. Publication of *Other Worlds* establishes the other side of Sagan's persona—as *fidei defensor* for traditional science and determined opponent of pseudoscience, in this case the theories of Immanuel Velikovsky (*Worlds in Collision*) and Erich von Däniken (*Chariots of the Gods?*).
1975–1985	Serves as a full member of Smithsonian Institution's council.
1975–1976	Serves as chair of the American Astronomical Society's division of planetary sciences.
1976–1996	David Duncan Professor of Astronomy and Space Sciences at Cornell University, a highly prestigious and visible professorship that he will hold for the rest of his career.
1977	Publication of *The Dragons of Eden: Speculations on the Evolution of Human Intelligence*, where Sagan successfully branches out his science writing to include disciplines that have relatively little to do with astronomy. Begins his long collaboration with Ann Druyan on the Voyager Interstellar Record Project.
1977–1979	Serves as chair of NASA's Study Group on Machine Intelligence and Robotics.
1978	Awarded a Pulitzer Prize for *The Dragons of Eden*. Publication of *Murmurs of Earth: The Voyager Interstellar Record* (coauthored with Frank Drake, Ann Druyan,

	Jon Lomberg, and Timothy Ferris). Serves as science correspondent for ABC News' *20/20*.
1979	Father, Samuel, dies. Publication of *Broca's Brain: Reflections on the Romance of Science*, which deals with a wide range of scientific issues—focusing on astronomy, with forays into other fields (ranging from evolutionary biology to artificial intelligence)—discussed within the context of Sagan's own reflections on humanity and the universe.
1979–1996	Serves as president of the Planetary Society.
1980	Broadcast of PBS miniseries *Cosmos* and publication of accompanying book. The success of *Cosmos* cements Sagan's celebrity status, making him the best-known living scientist in the United States.
1981	Receives a Peabody Award for *Cosmos*. He and Linda Salzman are divorced, and he marries Ann Druyan.
1982	Mother, Rachel, dies. Daughter, Alexandra (Sasha), is born.
1982–1996	Serves as a Fellow of the Robotics Institute at Carnegie Mellon University.
1982–1996	Distinguished Visiting Scientist at the jet propulsion laboratory of the California Institute of Technology.
1984	Publication of *The Cold and the Dark: The World After Nuclear War* (with Paul R. Ehrlich, Donald Kennedy, and Walter Orr Roberts). By this point in his career, Sagan has already become an advocate for nuclear disarmament—playing an important role in coining the phrase "nuclear winter"—and environmental sustainability.
1985	Publication of *Comet* (coauthored with Ann Druyan) and *Contact: A Novel*. Of Sagan's twenty-six books, *Contact* is his only work of fiction.
1988–1996	Appointed co-chair of the Global Forum of Spiritual and Parliamentary Leaders on Human Survival.
1989	Publication of *A Path Where No Man Thought: Nuclear Winter and the End of the Arms Race* (coauthored with atmospheric scientist Richard Turco).
1991	Son, Samuel, is born.

1992	Publication of *Shadows of Forgotten Ancestors: A Search for Who We Are* (coauthored with Ann Druyan). This marks Sagan's second foray into the evolution of humanity, following in the footsteps of 1977's Pulitzer Prize-winning *The Dragons of Eden*.
1994	Publication of *Pale Blue Dot: A Vision of the Human Future in Space*. Although Sagan had often written on environmental sustainability, it is in this book that he emphasizes the profound vulnerability of Earth and the need to protect it for the sake of humanity. Diagnosed with myelodysplasia, the rare bone marrow disease that will take his life.
1995	Publication of *The Demon-Haunted World: Science as a Candle in the Dark*, Sagan's book-length defense of skepticism and critique of pseudoscience.
1996	20 December: Carl Sagan dies of pneumonia, a result of myelodysplasia.
1997	Publication of *Billions and Billions: Thoughts on Life and Death at the Brink of the Millennium*. Major motion picture release of the film adaptation of *Contact: A Novel*, starring Jodie Foster.

Conversations with Carl Sagan

A Conversation with Carl Sagan
Timothy Ferris / 1973

From *Rolling Stone*, June 7, 1973. © 1973 Rolling Stone. All rights reserved. Reprinted by permission.

At a time when there is so much provocative nonsense around, it's nice to encounter some provocative sense. Carl Sagan's reputation as a brilliant astronomer with a gift for plain talk began around 1965 with the appearance of *Intelligent Life in the Universe*, a book he co-authored with the Soviet astronomer Iosef Shklovskii. Though never a best seller, *Intelligent Life* soon came to be known as one of the most exciting nontechnical science books ever written.

In a burst of energy following his work on the Mariner project (which put a satellite full of cameras into orbit around Mars), Sagan has written or contributed to five more books, all due out this year. Most of these center on his speciality, exobiology, the emerging science that concerns itself with life beyond the earth.

Trained in astronomy, physics, biology and genetics, Sagan lives with his wife and three children in Ithaca, New York, where he is director of Cornell University's Laboratory for Planetary Studies. In that laboratory we sat down one snowy January morning to talk.

TF: I'd like to ask you about the way the exploration of Mars has been reported in the press. What was your reaction when the earlier Mariner flights were going past Mars and there was a whole raft of editorials and articles about . . .
CS: "The Dead Planet."

TF: ". . . the Dead Planet." "Now we know that there is no life on Mars," and so forth. And the most recent Mariner mission—which provided an extraordinary opportunity to observe climate on another planet—was widely described as a disappointment, because dust at first obscured the surface of Mars. Is it discouraging to you that this opportunity for people's consciousness

to be expanded has been treated by the press as an opportunity instead to do the opposite?
CS: Yeah, it is a disappointment. But I've thought about this precise business a lot and those early reports about "The Dead Planet" are kind of interesting. Their logic is the kind of logic nobody would use in any other area.

For example, *Mariner IV* flew by Mars on Bastille Day 1965 and got twenty pictures of the planet with the finest detail one kilometer across. Now you take twenty pictures of the Earth at one kilometer resolution, there's *no chance* of finding life here. If there were kilometer-long elephants cheek by jowl covering the entire planet, they would have been excluded. And yet people say, "Well, I didn't see anything alive on that planet, it must be a dead planet." What *terrible* logic. How come everybody's using that?

The *New York Times* in 1965 ran an editorial called "The Dead Planet," and the argument was that a magnetometer on board—which you know measures magnetic fields—didn't find any magnetic field, therefore the planet is geologically dead. Now we know from these pictures that Mars is not geologically dead. Then they went on to say geologically dead is dead, so there's no life on the planet; it's a lifeless planet.

TF: There seemed to be almost a passion to do it.
CS: Well, I think a keen insight into how a lot of people think about this was provided by Lyndon Johnson, who said—this is more or less an exact quote—"As one of that generation of Americans who had the pants scared off of them by that Orson Welles invasion from Mars broadcast in 1938, I'm glad to hear that there isn't any life on Mars." I think Lyndon Johnson was speaking for many Americans then, as he may not have done on other issues.

Some people, at least, are disturbed about the idea that there might be life elsewhere; even simple forms of life. And the idea that there might be civilizations more advanced than ourselves elsewhere upsets a lot of people. I'm not a psychologist but I have spoken to a lot of people on the subject, and I think that there is a sense of "let's keep the idea of where we are in the universe *tidy*." It gets very *complicated* if you imagine that we're only one kind of life where there are millions of other kinds, some of them much more advanced than us. That is precisely a mind-expanding experience, and some people are not interested in having their minds expanded.

I think it also bumps into people's religious prejudices. The sophisticated representatives of all the major religions have stated that there's no test of faith involved, that it expands the range of God's activities if he made life on other planets and all that. But still I think there is a kind of fundamentalist malaise about the idea of life elsewhere.

An opposite emotional predisposition also exists: People desperately *want* to believe there's life elsewhere. That comes up in a lot of the UFO enthusiasts and you can find it in a lot of eighteenth century popular writing on the planets, where every planet had a different kind of being: The Mercurians were mercurial, the Venusians were amorous, the Martians fought a lot and the Jovians were jolly.

It seems to me an important issue whether there is life elsewhere. On important issues, you shouldn't make a decision until you have the evidence. But some people find it difficult to withhold judgment until the data is in. Its unsettling. I once wrote a book for Time/Life, a popular book on planets, and I would say, "Here's the relevant data; some people think this is the explanation; some people think that is the explanation." The editors of *Life* would come back to me and say, "Look, don't confuse our readers with the alternatives; just tell us what's right." I would say, "I don't know what's right. There are several possibilities, and we have to withhold judgment." They would say, "Well, just pick *one*. Whichever you like the best." I have the feeling that the editors of *Life* are in keen attunement with the way a lot of people think—with an intolerance for ambiguity.

TF: Has there been any new evidence of the existence of planets of other stars? In your book you mentioned that Barnard's star, a red dwarf about six light years away, has been found to have a dark companion about half again as massive as Jupiter. You described this object as "almost certainly a planet."
CS: The Barnard star situation is interesting. What you have is a residual in the apparent motion of the star. That is, here's a nearby star and you can plot very accurately its position in the sky relative to more distant stars that aren't moving. It's close and it's moving fast, so it has a large apparent or what we call proper motion. Superimposed on that proper motion there are little wiggles which are difficult to measure but have been measured over a period of decades and are certainly there. Now, the wiggles are due to a dark companion or companions, gravitationally on one side of the star and then on the other side, pulling the star one way or another. As to how many companions there are and what orbits they're in and masses they have, there is a range of possible solutions.

The original solution referred to in *Intelligent Life in the Universe*, and proposed by Peter Van De Kamp of Swathmore Observatory, was a single dark planet about one and a half times the mass of Jupiter, in a highly stretched-out elliptical orbit. Now he finds that he can match the data a lot better if he assumes two planets in circular orbits, like the orbits of our planets. They have just about the mass of Jupiter, but they're in a bit closer to their star than ours is. If you wanted to assume, say, eleven planets, you could fit the data even better. The main point is not that he has uniquely found two versus one but that far and away the most likely explanation of this motion of Barnard's star is planets of roughly Jovian mass.

TF: Would someone, say at the distance of Barnard's star or the star Sirius, observing the sun with equipment similar to that which we have now, be able to observe perturbations of the sun's motion or in some other fashion discover the existence of our planets?
CS: That question in perspective is a very nice one. First of all, there's the question of what does our sun look like? We did a computer program here a while ago in which we gave the computer the positions of the nearest one thousand stars and then asked it to draw star maps from the position of each. Of course, the relative orientation of the stars changes, which is another way of saying the constellations are different. My wife and I had fun making up names of new constellations. You know, constellations are just psychological projective tests; you look up and say, "That reminds me of a bear, I'll call it The Bear."

The remarkable thing is that even from the nearest star the sun is extremely unspectacular. For example the constellation Cassiopeia is in our northern skies and it's a kind of "W." Well, if you were in the vicinity of Alpha Centauri, the nearest star to our little one, four light years away, and you looked in the direction of Cassiopeia, you would see a "W" OK, but then there would be a final jog down. There would be one more star there, just about as bright as any of the others in Cassiopeia: That's us. That's the sun.

You know, our sun looks just like thousands of other stars in the sky. You'd never guess that there are planets going around it, and that one of those planets has people who consider themselves very intelligent. There would be no way of knowing that.

Here on the Earth if you go and look up some clear night you can see a few thousand stars. How do you know that they don't all have planets and guys standing around thinking that *they* are the smartest guys in the universe?

As far as detecting the Earth by gravitational perturbations, even from the vantage point of the nearest stars you can't do it. The Earth is just not massive enough. It's just too insignificant a planet. You could probably detect Jupiter and Saturn from the distance of the nearest stars with techniques not much more advanced than what we have today. But you'd never detect the Earth gravitationally from that distance. And if you went to any greater distance, you would not even be able to detect Jupiter and Saturn.

TF: Are those star maps in existence?
CS: Yes. We're thinking of making a children's book with pictures called *The Sky from Elsewhere*.

TF: I want to ask you about the conference on communication with extraterrestrial intelligence you attended in Armenia.
CS: This is something that a couple of us and a couple of Soviet astronomers tried for some years to get organized. It's not very easy to have an interdisciplinary meeting on such a speculative subject that involves two nations as much at odds as the U.S. and the Soviet Union. So merely holding the meeting represented something of a victory.

We had astronomers, physicists, chemists, biologists, anthropologists, archeologists, linguists, historians and one or two people who I'd call philosophers, plus people in computer sciences and electrical engineering. It was a remarkably diverse group and the quality of people was extremely high. We met for about five days at the base of Mt. Ararat, on which Noah's Ark is said to be beached.

The main conclusion was that it is not unlikely there are civilizations in advance of our own elsewhere in the galaxy and that we have means currently at our disposal to detect them. This doesn't mean that the conference committed itself to guaranteeing the existence of extraterrestrial intelligence, just that we cannot exclude the possibility. Some people think it's likely, some people think it's not very likely, but nobody can *exclude* it.

The Russians announced that for the last four years they have been doing a small project to examine the closest stars which are like the sun at two frequencies in the radio spectrum to see if there are any intelligible signals. The answer so far has been no. Even though that's a reasonably modest program, I think it is of interest that the Soviets have made such a sustained effort.

The thing that impresses me is that we have a capability with existing radio-telescopes for tuning in to an enormous number of stars, and we're not doing anything at all in the United States. For example, the world's largest semi-steerable telescope is Cornell's Arecibo Observatory in Puerto Rico. It's getting resurfaced, and it has a set of new receiving equipment.

Let's imagine Arecibo used 1 percent of its time to listen for some other civilization's signals, and imagine there to be another civilization just at our level of development, so they also have an Arecibo instrument to use as a transmitter. How far away could that other Arecibo be for us to detect it? The answer is that except for obscuration and dust in certain places it could be *anywhere* in the galaxy, and we would pick up the signals. That means at least one hundred *billion* stars that you can listen to for signs of extraterrestrial intelligence. In the United States we've listened to just two, back in 1960. The Russians have listened to something like a dozen.

So the situation is not that we have to build some vast and expensive new instrument to listen. We already have at hand the instrumentation necessary to muster such a search, and we're not utilizing it.

TF: When you put it that way, it seems astonishing that we're not.
CS: That's right. I'm hopeful that in the next few years the situation will turn around, and astronomers will be willing to spend a small fraction of their time on a regular basis searching for signs of extraterrestrial intelligence.

However, it's likely to be a very long search. You can't expect that you're going to go out and spend a few weeks and find it, because even under optimistic assumptions only something like one in one hundred thousand stars should have a civilization that we can communicate with. It may be much less than that, but I don't know anybody who thinks the chances are much better than one in one hundred thousand stars. So you've got to look at one hundred thousand stars, under optimistic assumptions, to have a good chance of picking one up.

TF: In your book you talk about how inconspicuous the Earth would seem from other stars. In Isaac Asimov's phrase, the solar system consists of the Jovian planets—Jupiter, Saturn, Uranus, Neptune—and debris. We are part of the debris; if you looked at the solar system from another star you wouldn't even notice us. Except, you point out, if you used a radio telescope, because ordinary radio and TV broadcasts in the past thirty years have suddenly made the Earth, in radio wavelengths, "brighter" than the sun itself.

CS: Well, the enormous amount of radio energy that we're pouring out today is due to three sources. One is the high frequency end of the AM broadcast band, another is just ordinary domestic television, the third is the radar defense networks in the United States and the Soviet Union. Those are the only signs of intelligent life detectable on Earth from a distance. It's pretty sobering. It's often asked, if there is extraterrestrial intelligence how come they don't come here? Now we know. Just listen to what we're sending out.

TF: There's a lot of soul music up at that end of the AM band though.

CS: Yes, and WQXR is at that end. There's a wide range of things at that end of the radio spectrum. But television and radar are the dominant thing.

Anyway, only for a brief moment in Earth's history have we had broadcasting. We're now going to cable television, the reason being that broadcasting wastes all that energy out into space when you're trying to talk to people on this planet. So soon we may be sending it all along various pipelines with nothing leaking out. And I could even imagine, if we don't destroy ourselves, our living with each other sufficiently happily that we are no longer constantly scanning for each other's missiles. Therefore it's possible that advanced civilizations don't leak out any radio energy.

It is much harder to detect the leakage that a civilization uses for its own purposes than it is to detect a signal that they are aiming at us for us to detect. When I was talking about there being a hundred billion stars within range of our hearing, that was under the assumption that some of those stars are sending a signal in our direction. If none of them are sending to us and they're just talking to themselves, then it is necessary to construct a very large array of radio telescopes in order to pick them up. That's called eavesdropping.

But remember, we're using a set of very restrictive conditions—namely that those guys are only transmitting as much power as we can transmit. We are the baby civilization in the galaxy, because we've just developed radio techniques in the last few decades. It's not likely that anybody else we can communicate with would be that backward. So anyone we can tune in on must be much smarter than us and therefore much more capable.

TF: So much of what you do necessarily involves such lengthy chains of speculation that it seems to me almost impossible to talk about it—these things that you've spent so much of your time working on—without employing suppositions so buried in our own civilization that we can't uproot them. A phrase that you used in another context was "assumptions intimately woven into the fabric of our thinking."

Just in talking about civilizations having progressed beyond ours, we may be victims of such assumptions: J. B. Bury wrote a book called *The Idea of Progress*, the thesis of which is that the whole concept of progress has existed only within the last couple of centuries of human thought. You suggest at one point that technological civilization itself may prove to be only a fleeting manifestation of intelligent life, possibly because it tends to quickly destroy itself. Doesn't it keep you awake at times just trying to trace some of these threads back, trying to get your thinking onto as solid a foundation as you can?

CS: Yes. It's a very important issue. I don't spend most of my time on these issues, largely for the reason that you've just very well stated—because it's not experimentally well-based yet. It remains in a very speculative arena. I spend some fraction of my time trying to make people aware that this is a very important question, but I don't pretend that the issue is solved at all. I think it's perfectly possible that there are few or conceivably no other civilizations in our entire galaxy of 250 billion stars. It's not out of the question at all.

But I can't imagine a more important scientific question, and we have in our hands the tools to approach it. I just can't understand why we're *not* doing it.

The general question that you ask is in the area which I like to call chauvinism. There's carbon chauvinism, water chauvinism—you know, people who say that life elsewhere can only be based on the same chemical assumptions as we are. Well, maybe that's right. But because the guys making that statement are based on carbon and water, I'm a little suspicious. If they were based on something else I'd give much more credence to it.

I must confess I'm a carbon chauvinist. Having gone through the alternative possibilities, I find that carbon is much better suited for making complex molecules, and much more abundant than the other things that you might think of. The standard science-fiction business of silicon replacing carbon just doesn't work well at all. The only circumstances in which it works are circumstances in which there is much more carbon, and so it always comes out second. I'm not that much of a water chauvinist. I can imagine ammonia, or mixtures of hydrocarbons, which are not all that cosmically rare, playing the role of water.

Then there are G-spectral-type chauvinists, who say that you can only have life around stars that are like our own; most stars are very different from our own. Planetary biology chauvinists say that life can only happen on planets, not for example in stars or in the interstellar medium. I'm a planetary biology chauvinist; there seem to be good reasons why life can only happen on planets.

The extreme chauvinist says, "If my grandmother would be uncomfortable in that environment, then life there is impossible." You come upon that pretty often. The phrase that you hear a lot, "life as we know it," is based exactly on that. It depends on who the "we" is. There are many exotic microorganisms on the earth which do well in solutions of hot concentrated sulphuric acid, and so on. If you don't know about them, you figure nobody could live in such an environment, but there are bugs that love it.

I think one of the great delights of exobiology is that it forces us to confront the provincialism in our assumptions about biology. All life on earth is essentially the same; chemically we're identical to bacteria or begonias. It's as though you said to a physicist, "You're going to study gravity now, but you can't go out of this room, and you can't look at anything that has a gravitational influence except what's within this room. Here are two big lead spheres. Measure how much they attract each other and try to devise a general theory." Well, that's very difficult. Newton did it not by being in a laboratory, but by looking at the motion of our moon and the moons of Jupiter and so on, and things on the earth as well. By making those connections he was able to make a general law of gravitation. Well, the biologists have mighty few general laws, and that's because they have mighty few cases—like one.

TF: When you look at speculation about possible other forms of life, it seems to me a lot of it is on one hand simply too fanciful. You mentioned, for example, that huge creatures with bone skeletons on an earth-like planet cannot exist because beyond a certain size bones don't have the strength to do it. The skeleton would have to be steel. And enormous insects in an earth-like environment are likely to exist only in human fantasy, because insects breathe by virtue of diffusion, which is not efficient enough to keep a big creature alive.
CS: That's why motion pictures like *Mothra* are flawed.

TF: Haven't seen *Mothra*.
CS: I haven't either, but I understand it's a very large insect. Maybe I'm wrong. If it's not a large insect then I don't object to it.

TF: So on one hand, speculation can be too fanciful. On the other hand a lot of it is too conservative—that's the consequence of limited imagination. In what ways can people on either side open up their thinking and realize the great variety of real possibilities without just slipping over into opposite errors?
CS: The only way is experimental. I just don't think you can sit down and think and get rid of all that accumulation of prejudices and fantasies. The way our minds think is the result of millions of years of evolution—hunting and gathering food, shinnying up trees, mating, building fires and all the rest of it. The way we think hasn't been optimized for confronting intelligence elsewhere, because we've never had to. So I just don't expect that we're going to make much progress by pure thought. The way we make the progress is to make the confrontation. Let's get the extraterrestrial message and then very carefully and very slowly try to come to grips with it.

The first part of your question brought to mind Mars. Mars has had the fanciful elements—Lowell's canals and all that—and it's had the almost fussy chauvinistic approach—"Oh, it's just like the moon." That second argument went, "The moon has craters; the moon is lifeless. Mars has craters therefore Mars is lifeless." Aristotle would turn over in his grave if he heard a syllogism like that.

Well, what's the reality as revealed by *Mariner IX*? We don't see any canals with liquid water running through them, but we do see things that look for all the world like dried-up rivers. We don't see a planet that's like the moon either; we see something just *different*. It's just fantastically *different* than anybody guessed. And I think that's where the reality is going to be in the search for extraterrestrial intelligence. It's not going to conform to our fantasies, and it's not going to conform to our chauvinism.

TF: Is the way in which scientists view science as a discipline changing? Charles Whitney has a book out called *The Discovery of Our Galaxy*, and he says in the very last sentence . . . "Scientists are releasing themselves from the strait-jacket of purely rational analysis. Some have come to view themselves as poets attempting to test their poems, or something along those lines."

Is the scientific method changing from a purely deductive, rational method to a more creative activity which tests itself against coherent data? And is our conception of the universe changing from seeing it as entropic and random to a view that it is essentially unified, and that the things that science treats are only part of a greater unity? Is that actually going on?

CS: I don't think science has ever been all that deductive. The cutting edge of science is always following the wild hunches, tracking out the clues, and that sort of thing. It differs from art only in that it makes a different sort of confrontation with reality. There's a test of whether a scientific theory is right or wrong: Namely, does it correctly predict all the things I can measure? That's different from the test of the success of a work of art. But I think that the kind of internal excitement motivating the scientist is very artistic. It's that same kind of searching for order and meaning, a quest for how the universe is put together.

I think we're constrained in how far we can go. Not by the scientific method; it seems to me the only reasonable approach, the one that confronts the data. Otherwise how would you ever know if a view were right or wrong? I think we're constrained by our minds. For the reason I mentioned before, our minds are put together the way they are because of the needs of a very different sort of existence in which human beings evolved—a hunter-gatherer society—and now we're asking that sort of brain to approach quite different circumstances.

It's remarkable that it does as well as it does. The thing that I find astonishing is that we are able to invent simple rules and constructs which are able to predict quantitatively a wide range of natural phenomena. I mean, how is it that we can have one little simple equation which describes pretty closely how bodies fall, no matter where on earth they fall or where you throw them or what their shapes are. You know, it's just a couple of little equations which are taught in high-school physics. Why is the world put together in such a way that we are able to construct these little equations which explain such a wide variety of phenomena? That's the astonishing thing.

The answer to that may be merely that things falling were pretty important to our ancestors, who lived in trees or something, so our minds evolved in such a way that things falling was something we had to understand. Those guys that couldn't understand it all fell out of the trees and broke their necks. We're not their descendants. We're the descendants of the guys who could understand how things fell.

But on the other hand, understanding how things fall here gives us a law of gravitation which happens to describe how two galaxies orbit each other. *That's* pretty fascinating. . . . Einstein said he found that the most astonishing thing of all is that we're able to understand as much as we can. He was not astonished that there were some things that we couldn't understand; that of course is what you'd expect.

Now there is a kind of dichotomy that a lot of people draw between rational and mystical. I'm not sure that that's a real dichotomy. For example, the thing that's described under the drug experience is to be one with the universe. Of course it's also described in non-drug religious experience. Eastern religions, Christianity, all have something like it. If you ask somebody who's had such experiences what he means by "one with the universe," well, of course there's great difficulty in converting it into words because it's a highly nonverbal experience. But I haven't found anybody who, while having that experience, was able to test it out.

You know, "Terrific, here I am, one with the universe, now I'm going to ask a question which nobody on the earth knows." OK? In detail. And come out of the experience, and say, well, "That was really a fabulous experience, and incidentally if you will perform the following experiment, with deuterons into a vanadium target, you're going to get the following result." Everybody says nonsense, but you do it, and it turns out that you were right. If that happened then I'd be much more willing to believe we were synched up to something we don't now understand.

So while not at all taking away from the ecstasy of such an experience, I'm skeptical about whether it really makes contact with the way in which the universe is put together. I think it makes contact with the way our skulls are put together, which is a different thing. I think mystical experiences may be excellent ways to find out about ourselves, at least it seems to me that would be the case, but I don't think we find out what's outside of ourselves that way.

The so-called rational approach seems to me, for all its shortcomings, to be the only way that works. I don't mean by that to justify at all the kind of mindless rationalism in which people say, "Don't ask me what happens to these poison gas canisters, I'm just doing my job." That's not what I'm talking about at all. Rationalism is not the suspension of ethical judgments. We're talking about finding out about the universe. I think that the rational approach, or if you want to call it that, the scientific method, is the way to go. The thing I stress is that it's driven by strong emotions. It's not dispassionate.

Timothy Ferris / 1973

The scientist—the real scientist, not the drudge—is a guy who is strongly motivated to find out about things around him and who would do it even if he wasn't making some money or recognition off it.

TF: Have you been reading more or have you turned up more evidence about the possibility that we have already been involved in interstellar communications? There is an argument that recorded history may make up only a small and not particularly important portion of man's real history. William Irwin Thompson suggests that, as he puts it, "Something has been communicating with us through the epochs of our civilization." He holds to the idea that communication historically could have been going on over periods so long that only mythology provides a vessel durable enough to accumulate any of the information.

In your book you include material gathered from Sumerian archeology. And of course there are many legends other than the Sumerian which are striking because you find such widely separated civilizations sharing a seemingly common mythology—a belief that civilization was derived from some high order of beings who passed it along to some sort of a priestly class of people and then disappeared. So my question is whether that seems to you a profitable avenue to pursue.

CS: Yes, I pursue that because one, it's a logical possibility, and two, it seems silly to spend a lot of money looking for life elsewhere if we have the evidence right here on earth. The conclusion I came to is that you'll never prove anything by legend alone. There are just too many possibilities, even with very similar legends, there are two classic possible explanations. One is that they in fact had contact among themselves. There was a huge amount of cultural diffusion in primitive time; even though it took a long time to traverse from Europe to Asia, those traverses were being made.

Secondly, there's a possibility that some kinds of things are wired into us. After all, birds have wired into them how to build nests, fly south for the winter and so on. There may be certain images that are wired into human beings, in our genetic material. Therefore human beings in very different places may have similarities in their thinking. I don't consider that a bizarre idea at all.

The only situation in which such a legend would be believable would be if it was remarkably detailed. The gods gave us information, and we didn't understand what they were talking about, but in the thirteenth century Irish monks copied it down, and in the sixteenth century somebody cataloged it and noted what was in there but he didn't understand what it was about either; and now it turns out to be details for the construction of a transistor radio. Well, such a legend I'd be willing to consider extremely seriously. But it's never anything like that. It's, "They came down and taught us how to write, do agriculture and regulate our behavior." That seems to me to have many other possible explanations.

The other possibility of course is to find the artifact, to find a sample of extraterrestrial technology that could not have been created by human beings because we weren't technologically up to snuff at that time. Those two cases I would certainly consider worth paying a lot of attention to. But the usual sort of legends about beings that lived in the sky and were not human beings—there are just too many other ways of understanding that for me to think they are serious clues to extraterrestrial intelligence.

TF: Is it realistic to think that there might not only be planets on which the environment is too hostile to life to exist, but also planets on which life is too comfortable for intelligence to derive? I'm thinking of the suggestion that the Ice Age may have had to do with the genesis of civilization on earth, and on the other side Arthur C. Clarke's suggestion that the chief difference between men and dolphins may be simply that at a point in evolution, the ancestors of the dolphin turned around and went back into the ocean and we didn't. They seem to be having a good time, and we have civilization. Is that a meaningful question?

CS: Oh, yes, it's a meaningful question. Unfortunately there are no meaningful answers. The Ice Age suggestion is Toynbee's idea of challenge and response. I think there is some aspect in which that is right. But as to the question of what are the accidental rare factors necessary to make an intelligent being by slow evolutionary process, and what are the factors that develop a civilization, nobody knows. The reason nobody knows is first that you can't do experiments on it—apart from ethical questions it would just take too long—and secondly, the one technical civilization that's developed on this planet has the awkward tendency to wipe out all the other civilizations that haven't yet achieved technical expertise. We never find out what would have happened to Aztec civilization if we had left it alone.

TF: If that tendency proved to be inherent in technological civilizations generally, it would be an awfully good argument for our not broadcasting anything.

CS: Yeah, this argument comes up many times. The main point to bear in mind there is that it's too late. We've already broadcast. A wave front of electromagnetic radiation is spreading out from the Earth at the speed of light and contained in it are arias by Enrico Caruso, the 1924 election returns, the Scopes Trial. . . . We've sent it, to say nothing of other things we've been talking about like television and the semi-paranoid radar defense networks of the major technical powers. So, it's just too late to say we shouldn't send. We have sent.

But my guess is that's not where it's at. The spaces between the stars are just enormous, and it's so difficult to do an interstellar journey that we cannot pose any threat to another civilization hundreds of light years away, and that distance between civilizations is an optimistic assumption. Even at the speed of light it would take hundreds of years to get there, and we certainly can't travel anything close to the speed of light. So there's a kind of imposed quarantine, at least at our level of civilization. There is no way in which we can pose a threat to any other civilization *and they must know it.*

As for the other kind of paranoid fantasies—that they'll find out we're here and come and eat us because we're so tasty or something—that doesn't work because the freightage is too expensive. If you found human beings had a particularly tasty sequence of amino acids in their proteins, you'd take home one human being and synthesize the protein and artificially mass-produce it. The gourmets on some other planet would then eat stuff that was made on that planet.

No, I think that this is the result of just not thinking the implications through carefully enough. I don't think anybody poses that kind of threat to us, and the sort of threat we might pose to somebody else is constrained by the vast distances between the stars. Also, mankind is getting better.

TF: One thing about those vast distances, though, is that in your writing it sometimes begins to seem easy to traverse them. For example, your mentioned that if we could build a spaceship able to maintain an acceleration close to the force of gravity on earth, we could travel to the center of the galaxy—thirty thousand light years away—in only twenty-one years measured on board the ship, owing to Einstein's time dilation.
CS: But that's for a technology that we're nowhere near obtaining, if it's possible at all. That starship capable of accelerating at 1 g reaches 99 percent the speed of light as time goes on. It never reaches the speed of light because of

the fundamental restrictions of special relativity. We must be at least centuries away from having such devices.

For example, a spacecraft called *Pioneer F* is on its way to Jupiter. It will get a big acceleration when it passes Jupiter, like the whip at the county fair, so much so that it will become the first man-made object to leave the solar system. At the speed it's going, how long before it gets to the distance of the nearest star? About a hundred thousand years.

That serves to calibrate the difference between the kind of 1 g constant acceleration starship that I was talking about and where we really are at. We don't know if it is possible to have such a 1 g starship. But if it is, we're certainly a huge way from having it. Some other guys may have it, but even if they do, I suspect it's extremely expensive and they don't just go tooling around for Sunday drives.

TF: It appears that we're also hundreds of years away from holding a dialogue with another civilization. You estimate that the average distance between intelligent civilizations, based upon a chain of conjecture, is maybe a hundred to a thousand light years.
CS: Yeah, say three hundred light years. So that means that they send a signal that says "Hello, how are you?" and we send back saying "Fine, thank you," and that takes six hundred years or something.

TF: Six hundred years. That would be. . .
CS: Thomas Aquinas's mother.

A Resonance with Something Alive
Henry S. F. Cooper, Jr. / 1976

From *The Search for Life on Mars*, Henry Holt and Company, 1979. Reprinted by permission of Henry S. F. Cooper, Jr.

On clear nights, in high, remote areas, Mars is seen to glow with a steady, hard reddish light—something it rarely seems to do from the city, through whose smoggy air the planet looks wan and bloodless and is difficult to identify. Right now, Mars is on the far side of the sun, and thus in the daytime sky, so that it cannot be seen at all. Normally, though, the planet can be picked out because it is brighter than the stars around it, and, unlike them, it doesn't twinkle. Through a telescope, Mars is not a pinpoint of light, the way all stars appear through even the most powerful instruments, but a round reddish ball—clearly a place, like our own globe, to which one might travel. In a number of cultures, the red planet has been associated with war, and in the last hundred years it has been the battle-ground of a scientific one: a dispute over whether life exists there. Of all the planets in the solar system, aside from our own, it is considered the most likely to harbor living things. With this possibility very much in mind, the National Aeronautics and Space Administration, beginning in the mid-sixties, has sent three spacecraft to observe the planet by flying close by it, and a fourth into orbit around it. Meanwhile, between 1962 and 1973, the Soviet Union dispatched eight spacecraft that we know of to Mars. Three of them vanished, while five sent back data with varying degrees of success. This summer, two more American spacecraft, designated *Viking 1* and *Viking 2*—each composed of a lander, which will descend to the surface, and an orbiter, which will continue to fly above it—will reach Mars; through a curious conjunction of celestial mechanics and more worldly concerns, the first lander is scheduled to touch down on July 4th, the day of our Bicentennial.

Currently, the most ardent advocate of the possibility of life on Mars, and on a lot of other places distant from our planet, is Dr. Carl Sagan, a professor of astronomy at Cornell, who has been on the scientific teams planning several of NASA's unmanned spacecraft missions. On August 10 last year, the day before *Viking 1* was supposed to be launched from the Kennedy Space Center, in Cape Canaveral, Florida, Sagan was addressing a dozen or so children seated on the hot cement near the pool of the Ramada Inn at Cocoa Beach, about twelve miles from the launching pad. A youthful-looking man of forty-one, with long, straight black hair combed at a sloping angle across a high forehead, Sagan (who pronounces his name to rhyme with "pagan") is a controversial figure, but most scientists will agree that if he doesn't embody the spirit of the whole Viking enterprise he at least supplies its imagination. On this occasion, he was dressed in black bathing trunks and a maroon-and-white patterned shirt and was sitting at the edge of the pool. At his feet were a partly broken model of a Viking lander, squat and froglike, and also, cradled in a wastebasket, a ruddy-colored globe of Mars as big as a beach ball. The children, most of them under ten, were sons or daughters of Viking scientists or engineers—Sagan's four-year-old son, Nicholas, was among them. They seemed to like Sagan, a man whose own childhood never seems very far behind him; he has remained close to it and seems to draw from it a rich and playful imagery.

"Here on earth, we have pools and beach balls and hot dogs and other nice things, but if you were far away you wouldn't see these things," he began. "From space, the earth would be a blue dot among lots of other dots—blue ones, green ones, brown ones, and, in particular, one big red one." He picked up the globe of Mars. "There is snow up here and down here," he said, touching the poles, where two holes had been punched so that the globe could revolve on a stand. "These holes don't exist on the real planet. But there's a giant mountain here. And here there's another. And here's another. And here's a huge canyon that would stretch from New York to beyond San Francisco if it were on earth. We know the mountains and the canyon exist because we can see them from space near Mars, but what we don't know is what is down *on* Mars." Though he didn't actually say it, he left the impression that if there actually were no hot dogs on Mars there might well be manifestations of life almost as interesting. One of Sagan's favorite arguments is that if a few thousand years ago, before there were advanced civilizations on earth, a spaceman from another planet had had a view of the earth no better than the one we have of Mars he might not know that any living thing was here. "So we want to send someone to Mars to see what's there," he went on to the children, who all looked very blond, and most of whom had recently come from such places as Hampton, Virginia; Denver, Colorado; or Pasadena and Mountain View, California—the sites of one or another of the factories, universities, and space centers where work on Viking had been going on. "We thought of sending a Martian, but we didn't know any Martians. That's one reason we're going. We asked some of our friends if they could live on Mars, but none of them could. It's too cold and too dry, and the atmosphere is mostly carbon dioxide instead of oxygen, like ours, and is only about one-hundredth the density of our own. So we had to make a person, and his name is Viking." Sagan picked up the model of the spacecraft. "He's a very special guy, and now I'll tell you what he has. He has three feet; he can't walk on them, but he can bounce a little as he lands. He has one giant ear inside his belly; it couldn't hear you, but it could hear an earthquake—a Marsquake—a thousand miles away. He has two ears on top, which turn, to hear radio signals, and he can talk with them, too. He has two eyes, like ours, only they're on stalks, like a crab's; he can see all the same colors we can and some that we can't. Now let's talk about mouths. He has three mouths, one of which is also a nose. With them, he eats dirt."

"*Yuk!*" said a girl with a very blond ponytail dangling down a very bronzed back.

"Yuk for you, but he likes it," Sagan said. "He doesn't eat for energy—he has all the food he needs inside him. He eats because he likes to. He has especially good tasters to tell one kind of dirt from another. He could easily smell the chlorine they're putting in the pool. He could even tell if there was anything *alive* in what he was eating. He has a hand to pick up the dirt he eats. He can pick up other things to look at. He has two arms. One is ten feet long, so he could reach over and pick up that girl over there. The other is shorter, and he uses it to feel the air. It's very thin air. Every day, he radios the Earth and says things like, 'Hello, Earth. The temperature on Mars is seventy below zero. It is very windy, and we don't have any snow.' He's pretty smart. He has a vocabulary of eighteen thousand words. An eight-year-old child knows perhaps half that many."

"Would you rather have Viking or an eight-year-old on Mars?" asked a boy in red bathing trunks who looked about eight.

"I'd rather have an eight-year-old," Sagan said at once. "Viking may be smart, but he's slow. If a fat Martian walks by, Dr. Anderson, the scientist in charge of the big ear, will go to Dr. Mutch, the scientist in charge of the two eyes, and say, 'I hear something fat walking around out there.' Then

Dr. Mutch will go to Mission Control and say, 'There's something fat walking around out there. Let's look for it.' *Three days later* Viking looks for it, and by that time whatever it was will have lumbered out of view. Another thing that Viking can't do is reproduce. It would be nice if these two Viking landers—there are two of them, remember, each with its own orbiter overhead—could make a lot more, but they can't. Anyway, that's our special guy on Mars, and tomorrow he'll go off. This is the first time something will actually land on Mars and tell us about it, so you're very lucky."

"What if it blows up?" the boy in red trunks asked.

"That's one reason we have two of them," Sagan said.

"What if a leg falls of?" the girl with the ponytail asked.

This had happened to the model that Sagan had in his hand.

"Then it will have a limp—Viking will tilt," Sagan said.

"What if a Martian cuts off an eye?" another boy asked.

"That will be terrific! Then the other eye will see him."

"What if the Martians have sophisticated weapons that blow it up?" another girl asked.

"Then we'll have a blown-up lander," Sagan said. "And maybe the cameras will photograph the Martian doing these bad things. But the Martians probably won't be bad. They will either be kindly or they won't care about us."

With his playfulness, his ability to bring science fiction to the aid of science, and his nimble way of turning a question inside out, so that an adverse circumstance suddenly becomes an asset, Sagan alternately delights and infuriates not only children but his scientific colleagues as well. The latter don't know quite what to make of him, for although they regard him as a good, even brilliant scientist, they have trouble coming to grips with his most distinctive quality, his imagination. Sagan is a theorist—a type of scientist who traditionally irritates many of his fellows, because he necessarily deals with what might be instead of what is. Scientists can be quite tough on colleagues who they feel speculate too much, especially in public. Sagan believes that the most important question facing mankind today is whether there is life, intelligent or not, elsewhere in the universe; although the subject is one that is full of pitfalls, he and a large part of the scientific community have in recent years come to feel that there must be such life. As yet, there is no direct evidence for extraterrestrial life, although one biologist who is a good friend of Sagan's, Dr. Joshua Lederberg, of Stanford University, adduces as certain proof that there is life in space one impressive set of evidence: ourselves.

There are, however, two generally acknowledged areas of indirect evidence of exobiology. (Exobiology, meaning life outside this planet or the study of it, is a word that was coined some fifteen years ago by Lederberg, who, in conversations with Sagan and others, found extraterrestrial biology, from which it derives, too much of a jaw-breaker.) One area of indirect evidence is the vast number of stars in the sky: there are an estimated two hundred and fifty billion of them in our galaxy alone, and within sight of our largest telescopes there are probably as many galaxies as there are stars in this one. Since planets are thought to be a common consequence of star formation, a large proportion of these stars presumably have solar systems, though none except our own star are known to have. This line of reasoning would not prove much unless it could also be shown that life will occur where conditions are right, and the second area of indirect evidence suggests that it will. In the last couple of decades, molecular biologists, who are concerned with the formation of the molecules on which life is based, have demonstrated what they believe is the way life on earth developed from the simplest organic compounds (organic compounds are those based on carbon). Many of these compounds have recently been discovered in such profusion throughout space that many biologists are convinced that life not only is a common manifestation in the universe but may actually be an inevitable consequence of it. If there is no life elsewhere, Sagan says—turning the argument back on itself—then scientists will be faced with what he regards as a much more difficult problem: explaining what is so special about our particular part of the universe that life developed only here. Sagan likes to quote a friend of his—Dr. Philip Morrison, a physicist at the Massachusetts Institute of Technology, who is currently the chairman of a NASA committee to recommend methods of communicating by radio with extraterrestrial civilizations. Morrison has said that the discovery of any sort of life on Mars, however meagre, would immediately change life from a miracle into a statistic—initially of two. Indeed, Sagan and others see life in the universe as a sort of statistical pyramid, as it is on Earth, with the lower forms vastly out-numbering the higher; consequently, if a microbe is found on a relatively arid planet like Mars, Sagan feels, many people would be willing to make what he calls "the great leap" to the acceptance of belief in a cosmos populated fairly consistently with intelligent beings.

Sagan pursues the matter of extraterrestrial life not only in laboratories but also in classrooms, in books, and on television (where he will be a familiar sight this summer, after Viking lands on Mars). He has written thirteen books, including three popular ones: *Intelligent Life in the Universe* (1966), on which he collaborated with the Soviet astrophysicist I. S. Shklovskii; *Other Worlds* (1975); and *The Cosmic Connection* (1973), which is his best known. It is a literate account of the likelihood of our discovering extraterrestrial life of all sorts, from microbial to intelligent, and its approach ranges from scientific objectivity to lyricism. Sagan writes well. In his preface to *The Cosmic Connection*, he says:

> Even today, there are moments when what I do seems to me like an improbable, if unusually pleasant, dream: to be involved in the exploration of Venus, Mars, Jupiter, and Saturn; to try to duplicate the steps that led to the origin of life four billion years ago on an Earth very different from the one we know; to land instruments on Mars to search there for life; and perhaps to be engaged in a serious effort to communicate with other intelligent beings, if such there be, out there in the dark of the night sky.
>
> Had I been born fifty years earlier, I could have pursued none of these activities. They were then all figments of the speculative imagination. Had I been born fifty years later, I also could not have been involved in these efforts, except possibly the last, because fifty years from now the preliminary reconnaissance of the solar system, the search for life on Mars, and the study of the origin of life will have been completed. I think myself extraordinarily fortunate to be alive at the one moment in the history of mankind when such ventures are being undertaken.

At the time of the launching of *Viking 1*, Sagan wore two NASA badges, one identifying him as a scientist and the other as a correspondent for *Icarus*, a scientific journal he edits. It was as though he was having a hard time deciding whether he was a scientist or a writer. Sagan says he sometimes gets bored with the company of scientists; he brought to Florida for the Viking launch, in addition to his wife, Linda, who is an artist, and his son, a number of other artists and some writers—people he knows or has worked with, including Stewart Brand, the editor of *The Whole Earth Catalogue*, whose 1971 edition contains extracts from some of Sagan's lectures on the planets. Sagan has lately been working with Francis Ford Coppola, the movie producer and director, on a script for a science-fiction film for television, and had invited him along, but at the last minute Coppola had to go to Australia. Sagan is well on his way to becoming a cult figure. Young people are fascinated by the idea of extraterrestrial life, and Sagan is much in demand as a speaker, especially at colleges. With his black hair, dark skin, and deep-set eyes, he has dramatic good looks of a kind that seem to appeal strongly to the female segment of his audiences. Recently, during a lecture he was giving in Houston, when he raised the question of whether there was life on Mars, and added, with a smile, "And if we go there we might have to listen to equally boring speeches," a tremendous, unladylike sigh filled the auditorium. Sagan's schedule of public appearances is a busy one, to judge by one month's engagements in the spring of 1975. On April 6, he was in New York addressing a symposium of gifted children (he was one himself once) organized by the Department of Health, Education and Welfare. On April 9, 10, and 11, he delivered a series of lectures about life on Mars at the University of Pennsylvania. On April 14, he was in New York again, serving as a judge for the National Book Awards. On April 23, he flew to Pasadena for a meeting on extraterrestrial intelligence, and while he was there he appeared on the *Tonight Show*. On his way back to Cornell, he stopped at Denver in order to give a talk to engineers at Martin Marietta Aerospace, where the Viking landers were made. Some of his colleagues think he is becoming too much of a showman, and Sagan himself is wondering whether he spends too much time on the lecture-and-television circuit; he was alarmed when his son Nicholas told him he wanted to be two things when he grew up—"a daddy and a host," meaning by the latter the host of a TV talk show, such as he has frequently seen his father appear on.

Sagan is a member of what is known as the imaging team for the two Viking landers—the scientific group that will analyze the photographs sent back from the surface of Mars. He and Dr. James B. Pollack, of NASA's Ames Research Center, in Mountain View, California, are the only astronomers on that team, which is made up mostly of geologists. More important, Sagan is the only member to have a strong background in biology. Although there is a separate team of biologists, and there are three specialized instruments aboard each lander to detect microbes, Sagan believes that the cameras could well prove the most effective means of discovering life on Mars—on the principle that the surest way to discover life on earth is to open your eyes. The cameras, he feels, will make the fewest assumptions about what life on Mars is like; the three biology instruments, in which Martian soil will be cultured to see if anything will grow, will make a number of assumptions about such things as temperature, nutrients, wetness, and metabolism. The cameras will make only one, but that, of course, is a whopper: that life on Mars will be big enough to see. Sagan is about the only Viking scientist who accepts this possibility, and he told me recently that he will make it his principal duty to search the Viking photographs for visible signs of life. It is almost beyond his colleagues' wildest expectations to find a microbe on Mars, let alone anything larger. "Carl serves an important function at some risk to himself," his friend Dr. Morrison said at the conclusion of a press conference the day before the scheduled launching of the first Viking spacecraft. (Owing to numerous glitches—aerospace gremlins—it didn't get off the ground for almost two weeks after that; in fact, at one point there were so many mechanical problems with Viking that one eminent scientist called it a "screwed-up mess.") At the pre-launch press conference, a group of seventeen Viking scientists were asked for a show of hands on whether they believed there was life on Mars. At first no hands went up; then two or three were raised; and after about a minute there were eleven. Morrison, who witnessed this demonstration of uncertainty, cited it as evidence that most scientists felt a certain sympathy for Sagan's more open espousal of extraterrestrial life—as though he were their collective unconscious. Most of the scientists in the group, however, thought that the tentativeness of their hand raising expressed their attitude better than Sagan's eloquence did. Most of them are conducting experiments on Mars which are much more prosaic than the task of searching the Viking photographs for visible signs of life, and it is just possible that if Sagan isn't there to do that, no one else will.

Sagan was born in 1934 in the Bensonhurst section of Brooklyn, where his father was a cutter in a clothing factory. "It was during the Depression, and we were kind of poor," he said not long ago. "When I was very little, the basic thing for me was stars. When I was five years old, I could see them at whatever time bedtime was in winter, and they just didn't seem to belong in Brooklyn. The sun and the moon seemed perfectly right for Brooklyn, but the stars were different. I had the sense of something interesting, distant, strange about them. I asked people what the stars were, and I mostly got answers like 'They're lights in the sky, kid.' I could tell they were lights in the sky; that wasn't what I meant. After I got my first library card, I made a big expedition to the public-library branch on Eighty-sixth Street in Brooklyn. I had to take the streetcar; it was some big distance. I wanted a book on the stars. At first, there was some confusion; the librarian mentioned all kinds of books about Hollywood stars. I was embarrassed, so I didn't explain right away, but finally I got across what I wanted. They got me this book, and

I read it right there, because I wanted the answer." (Sagan was a precocious child. So is his son Nicholas, who resembles Sagan in many ways. Nicholas taught himself to read by the age of twenty-one months—a fact that his parents discovered when he began rattling off the road signs on a transcontinental car trip. Sagan is an indulgent father, and, on occasion, he can be an ingenious one; for example, on that trip, during which he dictated large sections of *The Cosmic Connection* into a tape recorder, he also played back a number of children's stories that he had taped earlier, which kept the boy occupied between road signs. Nicholas is indeed wise beyond his years; recently, when he was asked whether he believed there was life on Mars, he replied, "Maybe yes and maybe no.") Sagan continued, "The library book had this stunning, astonishing thing in it—that the stars were suns, just like our sun, so far away that they were only a twinkle of light. I didn't know how far away that was, because I didn't know mathematics, but I could tell only by thinking of how bright the sun is in the daytime and how dim a star is at night that the sun would have to be very far away to be just a twinkle, and the scale of the universe opened up for me.

"It must have been a year or two after this that I learned what the planets were. Then it seemed absolutely certain to me that if the stars were like the sun there must be planets around them. And they must have life on them. This was an old idea, of course. Christiaan Huygens, the Dutch astronomer, I found out later, had written about it in the 1670s. But I thought of it before I was eight. And once I reached that point, I got very interested in astronomy. I spent a lot of time working on distances, coordinates, and parallaxes.

"Then, when I was ten—I was at P.S. 101 in Brooklyn at the time—I came upon the Edgar Rice Burroughs novels about John Carter and his travels on Lowell's Mars. It was a world of ruined cities, planet-girdling canals, immense pumping stations—a feudal technological society. The people there were red, green, black, yellow, or white and some of them had removable heads, but basically they were *human*. I didn't realize then the chauvinism of making people on another planet like us; I simply devoured what seemed to me the riches of another planet's biology. Carter fell in love with a princess of the Kingdom of Helium, Dejah Thoris. It was very exciting, and I loved those books. They were full of new ideas. On Burroughs's Mars, there were two primary colors more than on Earth, and I would close my eyes and try to imagine them. I tried to imagine my way to Mars, the way Carter did: I would go into a vacant lot, spread my arms, and wish to be on Mars." Thirty-one years later, Sagan has taped up on the wall outside his Cornell office a map of Mars as Burroughs portrayed it, with Xs marking the spots where Carter landed. Recently, in his office, he showed a visitor, on a globe of Mars made from *Mariner 9* photographs, exactly where Carter would have come down. "Many an evening I spent in vacant lots, arms outstretched, *thinking* myself to that twinkling red place, but nothing happened. I tried all different kinds of wishing. Suddenly, it dawned on me that this was fiction; maybe there was some better way to get to Mars.

"This was toward the end of the second World War, and I heard about the V-2 rockets the Germans used to bomb England. There were occasional references in the papers to what rockets could do for space technology. I found some journals put out by a group called the British Interplanetary Society. It sounded nice. And gradually I realized that there *was* a way. I discovered that in 1939 the British Interplanetary Society published a study for a multistage rocket that could go to the moon. If the moon, then why not Mars?

"I didn't make a decision to pursue astronomy; rather, it just grabbed me, and I had no thought of escaping. But I didn't know that you could get paid for it. I thought I'd have to have some job I was temperamentally unsuited to, like door-to-door salesman, and then on weekends or at nights I could do astronomy. That's the way it was done in the fiction I read, in which space science was practiced by wealthy amateurs. Then, in my sophomore year in high school, my biology teacher (this was at Rahway High, because we had moved to New Jersey) told me he was pretty sure that Harvard paid Harlow Shapley a salary. That was a splendid day—when I began to suspect that if I tried hard I could do astronomy full time, not just part time.

"I had been receiving catalogues from various colleges, and I wanted one with good mathematics and physics. The University of Chicago sent me a booklet entitled 'If You Want an Education.' Inside was a picture of football players fighting on a field, and under it the caption 'If you want a school with good football, don't come to the University of Chicago.' Then there was a picture of some drunken kids, and the caption 'If you want a school with a good fraternity life, don't come to the University of Chicago.' It sounded like the place for me. The trouble was that it had no engineering school, and I wanted an education not only in astronomy and physics but also in rocket engineering. I went down to Princeton to ask Lyman Spitzer, the astronomer, his advice; he was involved in some early rocket studies. He told me that there was no reason an astronomer had to know every nut and bolt of a spacecraft in order to use it. Up until then, I had thought this was necessary—another holdover from the fiction I'd been reading, in which the rich amateur built his own spaceship. Now I realized that I could go to the University of Chicago, even though it had no engineering school. I applied, and entered in the fall of 1951. In the early 1950s, the University of Chicago was a very exciting place to be. It was strong in the humanities—which I wanted—but it was also very strong in the sciences. Enrico Fermi and Harold Urey were both there, in physics and in chemistry. And it had a superb astronomy department, which operated the Yerkes Observatory."

Sagan began to attract the attention of older scientists, many of them Nobel laureates, who, after experiences in a variety of fields, were beginning to think about extraterrestrial life. Back in Rahway for Christmas vacation during his freshman year, he met a young biologist—the nephew of a friend of his mother's—who was at Indiana University, working with Dr. H. J. Muller, who had won the 1946 Nobel Prize in Medicine and Physiology for the discovery that X-rays caused mutations in genes; Sagan was interested, because X-rays are produced by exploding stars—novas or supernovas—and Muller's discovery showed them to be a direct link between astronomy and the evolution of life. The young biologist told Sagan that Muller was now working full time on the origins of life. Later, back at the University of Chicago, Sagan wrote his new friend a letter; the friend showed it to Muller, who liked it and wrote to Sagan asking him to spend the summer of his freshman year working for him at Indiana. "Muller had me doing routine things, such as looking at fruit flies for new mutations," Sagan said. "But he ran a real research group, and for the first time I got a feeling of what scientific research was like. Moreover, Muller was interested not only in the origins of life but in the possibility of life elsewhere; he didn't think the idea was the least bit silly." Muller, of course, was by no means the first biologist to concern himself with extraterrestrial life; that distinction probably belongs to Alfred Russel Wallace, and the tradition continues with J. B. S. Haldane and Alexander Oparin. "Muller encouraged me to learn genetics," Sagan went on. "Later, he sustained me through years of studying biology and chemistry, which I had thought were far removed from my main interest, astronomy. I always kept in touch with him. A few years before his death, he gave me a book about space flight by Arthur C. Clarke, and inscribed it, 'Perhaps we'll meet someday on the tundras of Mars.' He died in 1967. In 1973, after *Mariner 9* had mapped Mars, I managed to get a crater there named for him." Sagan is a member of the Subcommittee on Martian Nomenclature of the International Astronomical Union, which now handles such matters; he also participated in naming craters after Lowell, Schiaparelli, and Edgar Rice Burroughs.

In the fall of his sophomore year, Sagan returned to the University of Chicago with a letter from Muller introducing him to Dr. Urey, who had won the 1934 Nobel Prize in Chemistry for the discovery of heavy hydrogen, and had gone on to the study of the origins of life. (This was the basis of his interest in lunar science, a field of which he is generally considered the modern-day founder.) Urey is a mild-mannered man with a benign intelligence. "He was extremely kind to me when I was an undergraduate," Sagan said. "I did an honors essay on how life began. It was very naïve, and I remember Urey's comment: 'This is the work of a very young man.' I had the idea that in one fell swoop I could understand the origins of life, though I had not had much chemistry or biology. It was an attempt to learn by doing. Some other people at Chicago were more effective at this than I was. It was a time of great excitement, for this was when Stanley Miller was doing his work, under Urey, on the origins of life. He had filled a flask with methane, ammonia, water, and hydrogen—things you would expect to find in the primitive atmosphere of a young planet—and then had passed an electrical discharge, like lightning, through it. The result was amino acids, the first step toward life. Miller had shown that the beginnings of life were not a matter of chance but could happen in any place where the conditions were right." (Recently, Sagan showed a visitor to his laboratory at Cornell a version of Miller's experiment, with further ramifications; the resulting organic compounds were a reddish-brown gunk that gave the same spectrum as the reddish-brown belts on Jupiter.) He went on, "Urey showed me through Miller's laboratory. Later, Miller was forced to defend his work before the University of Chicago's chemistry department. They didn't take it very seriously; they kept suggesting that he had been sloppy, leaving amino acids all over his laboratory. I was outraged that something as important as that could be received in such a hostile way. Urey was the only one who spoke up for him. He said, 'If God didn't create life this way, He certainly missed a good bet.'" Urey, who is now eighty-three, is a member of the Viking team that will be looking for organic compounds on Mars.

After getting his master's degree in physics, Sagan went on in 1956, to the University of Chicago's graduate school in astronomy, which was at

Williams Bay, Wisconsin. There he worked with Dr. Gerard Kuiper, a Dutch astronomer who at the time was the only full-time planetologist in the United States. Though Lowell had alienated the astronomical community, he wasn't solely responsible for planetary studies' having fallen into disrepute. With the advent of astrophysics, in the 1920s, astronomy had taken a more professional turn, which was one that led away from the planets; indeed, Kuiper, whom Sagan sees as providing a sort of link between the earlier planetary astronomers and what he calls "the present burgeoning time," had to start by studying the stars before he could go into planetary work.

In 1944 Kuiper had discovered that Titan, the largest moon of Saturn, had an atmosphere, and that the major constituent of Titan's atmosphere was methane—a discovery that impressed Sagan, who later suggested that life might be found there. Kuiper was responsible for the idea that there might be lichens on Mars, because he had found that the planet's spectrum was not inconsistent with them (though it was inconsistent with green plants). Sagan didn't think much of the idea because he doesn't think that specific terrestrial organisms would be duplicated in the course of Martian evolution. Nevertheless, he welcomed the suggestion, if only for its propaganda value, because he was already becoming interested in making the theory of extra-terrestrial life acceptable again. "Kuiper was a respected man, and if he said it was possible for *any* sort of life to exist on Mars, that was important," Sagan said. "It was a tremendous boost to exobiology." Sagan had spent the summer of 1956 with Kuiper at the McDonald Observatory in Fort Davis, Texas, and there he had his first opportunity to see what Mars looked like in a close opposition through a big telescope. "As it turned out, there were dust storms in both places—Mars and Texas," he said. "I didn't find any canals. I was satisfied just to be able to see light and dark markings. The seeing was poor, even through the eighty-two-inch telescope at McDonald. There Mars was, though shimmering, squashed, distorted. Then, for an instant, the atmosphere steadied, and I caught a glimpse of the southern polar cap. I saw no fine details. It was no big deal. I realized that the telescopic technique, while interesting, was limited: sitting under a blanket of air forty million miles from the target was not going to tell me much."

While Sagan was getting his doctorate under Kuiper, he married a young biologist (they had two sons, Dorion and Jeremy, both of whom are now in high school), and the couple moved from Williams Bay to Madison, Wisconsin, whose university atmosphere was more to his liking. It was during this period that he met Lederberg, who was then professor of genetics at the University of Wisconsin. Lederberg, who had just won his Nobel Prize, had a reputation for brilliance and inaccessibility. "He was an object of some consternation and fear," Sagan said. "Postdoctorates in biology were afraid to present papers lest he be in the audience and demolish their thesis with two questions. Then, one day, he called me up out of the blue and said he wanted to see me—said he was interested in extraterrestrial life. I was immensely flattered." One of the ideas that Lederberg wanted to talk about was a thought he had had some years before when he first read Oparin's work *The Origin of Life on Earth*. It had occurred to Lederberg that since 90 percent of the universe is made up of the same few atoms that are required for life on earth—the main ones being hydrogen, carbon, nitrogen, and oxygen—then there should be no discontinuity in the development of organic compounds in the course of the evolution of the universe; that is, that they should be as common anywhere else—even in the space between the stars—as they are here. His theory was subsequently borne out: in the next decade, radio astronomers found clouds in interstellar space containing forty or fifty varieties of organic compounds, such as formaldehyde, and it was discovered that some meteorites contain them, too. Lederberg wondered whether postulating the generation of amino acids on each planet by means of lightning in its atmosphere, in the manner that Stanley Miller had demonstrated, was necessary; they could be raining down on the planets all the time. "What I like about working with Lederberg is that things turn out well both for biology and for astronomy," Sagan said recently. The two scientists immediately took to each other. Sagan describes Lederberg as "a dry, stimulating, totally unfettered man who is willing to carry his ideas to their logical consequence, even though the prevailing wisdom says it's silly," and as "a natural resource that should be used widely." He soon regarded Lederberg as a mentor and a collaborator. "We stimulated each other's ideas," he said. "It was a pleasure to talk to him; we didn't either of us have to finish sentences. We could leapfrog through arguments—an efficient way of talking. Then and since, I've had dozens of ideas that I've been able to bounce off him. Many are jointly arrived at, so that neither of us knows which thought of them first."

Lederberg concurs in much of this. "Sagan has fired up some of my ideas, and I think I've helped fire up some of his," he said not long ago. "Back in the early days of our friendship, I think it was helpful to Sagan that an established biologist could entertain the same thoughts he did about extraterrestrial life."

In 1959 Lederberg headed a committee of the Space Science Board of the National Academy of Sciences to study ways of searching for life in space; he asked Sagan, who was then twenty-five, to be a member. The group included several men who would be important to Sagan later, including Wolf Vishniac, who was then at Yale, and who shared many of Sagan's and Lederberg's ideas. Sagan, Lederberg, and Vishniac (who was killed in 1973 in a fall in Antarctica, where he had gone in search of microbes that might be analogous to ones on Mars) participated in many other conferences on extraterrestrial life, most notably a symposium sponsored by the Space Science Board in 1964–65, whose proceedings were published in a book, *Biology and the Exploration of Mars*, which provided an important scientific under-pinning for Viking.

Sagan received his doctorate in astronomy and astrophysics from the University of Chicago in 1960; he went on to become a research fellow at the University of California at Berkeley, and then, at Lederberg's invitation, he spent a year as visiting assistant professor of genetics at the Stanford medical school, to which Lederberg had recently moved. From 1962 until 1968, Sagan held a joint appointment as astrophysicist at the Smithsonian Astrophysical Observatory in Cambridge, Massachusetts, and lecturer and later assistant professor of astronomy at Harvard. In 1968, he moved to Cornell, where he is currently not only the David Duncan Professor of Astronomy and Space Sciences but also director of the Laboratory for Planetary Studies. In addition to the commissions and conferences on extraterrestrial life, Sagan has participated in the work of an enormous number of boards and committees having to do with space exploration, including the groups that formulated the international procedures for sterilizing spacecraft and several committees for NASA—most notably, of course, the imaging teams of *Mariner 9* and Viking. He also became involved in the flights of *Pioneers 10* and *11*. *Pioneer 10* is now on its way out of the solar system and *Pioneer 11* will follow after it has flown by Saturn. When he realized that *Pioneers 10* and *11* would be the first man-made objects to leave the solar system, he had plaques, designed by himself and a Cornell colleague, Dr. Frank Drake, placed aboard. Besides indicating the time of the launch in relation to the history of our galaxy, they bore a sort of return address, in case they fell into the hands (or whatever) of extraterrestrial beings; this included the coordinates of the Earth in relation to a number of radio sources in the sky, along with pictures of those who had launched the craft—delineations of a nude man and woman, drawn by his present wife, Linda. Sagan still gets letters from people complaining about his sending smut into space, and he delights in the fact that, because the paperback edition of *The Cosmic Connection* bears a similar depiction on the cover, people who see it in bookstores think it's a dirty book. He himself was the first to realize, however, that, because the two *Pioneers* would probably never come close to another star, the plaques were less an attempt to communicate with other civilizations than an effort to communicate with our own, the message being to make people more accepting of the idea of extraterrestrial life, whose image Sagan is continually trying to improve. Indeed, if he hadn't been a scientist he might have made his fortune in public relations. For a conference on extraterrestrial life held in Soviet Armenia in 1971, of which he was co-chairman, Sagan, who is fond of plays on words, promoted the name CETI, pointing out that it was not only the acronym of Communication with Extra-Terrestrial Intelligence but a reference to Tau Ceti, the nearest star with characteristics like our own sun, and also the plural of *cetus*, the Latin name for the whale, an intelligent species that offers us the same problems of communication as an extraterrestrial one could; the multiple pun is one of the main things about the conference which people remember today.

Grandstand plays such as these, combined with Sagan's scientific achievements, have brought him a good deal of attention: in 1974, *Time* named him one of "Two Hundred Rising American Leaders," and he has been awarded both NASA's medal for exceptional scientific achievement and France's Prix Galabert, which was awarded to him "*pour son éminente contribution personnelle à la connaissance ainsi qu'à l'exploration des planètes.*" His scientific achievements include over two hundred papers, some of them written in collaboration with Lederberg, or with Cornell associates such as Joseph Veverka, Frank Drake, and Peter Gierasch, or with James Pollack, who had been a student of his at Harvard and is now his colleague on the Viking-lander imaging team. Some of the more imaginative ones concern such matters as how microbes might exist below the lunar surface; how life might exist in the clouds of Venus; how life might exist in isolated areas on Mars; how Martian micro-organisms might survive the rigors of the planet by existing a centimetre below the surface; how the climate of Venus, enveloped in a hot, heavy atmosphere of carbon dioxide, which acts like the glass of a greenhouse, could be improved by dropping a certain type of algae into the clouds, which would break down the carbon dioxide and cool the planet; and how the Martian climate (if it is determined that there is no indigenous life) could be improved by depositing colonies of dark-colored microbes on the ice caps which could multiply and melt them (if Martian life is found, so that one would not want to introduce terrestrial microbes, carbon black could be substituted, and he has even calculated the number of rocket ships that would be required to get it there in sufficient quantities). A large number of Sagan's scientific papers have to do with communicating by radio with intelligent life elsewhere in the cosmos, and currently he is investigating this possibility with Frank Drake. Drake is director of the radio telescope at Arecibo, Puerto Rico, the biggest in the world, which is operated by Cornell for the National Science Foundation; the two scientists listen, when they can get the time, for signals from intelligent life in outer space. "Sagan desperately wants to find life someplace, anyplace—on Mars, on Titan, in the solar system or outside it," one of his Viking colleagues said recently. "In all the divergent things he does, that is the unifying thread. I don't know why, but if you read his papers or listen to his speeches, even though they are on a wide variety of seemingly unrelated topics, there is always the question 'Is this or that phenomenon related to life?' People say, 'What a varied career he has had,' but everything he has done has had this one underlying purpose."

Sagan was asked the other day why he thought it was that he, and others, are so interested in trying to find life beyond the earth. "I think it's because human beings love to be alive, and we have an emotional resonance with something else alive, rather than with a molybdenum atom," he said. "Why are people interested in other animals? Why are we interested in the life history of the armadillo? Why do we go to Antarctica to find out what the emperor penguins have been doing lately? It's fun, because we are primarily drawn to things that are alive."

Carl Sagan Interviewed
Joseph Goodavage / 1976

Originally appeared in *Analog: Science Fiction and Fact*, August 1976, pp. 92–101. © 1976 Condé Nast. Used by permission.

"There is a place with four suns in the sky—red, white, blue and yellow; two of them are so close together that they touch, and star-stuff flows between them. I know of a world with a million moons. I know of a sun the size of the Earth— and made of diamond. There are atomic nuclei a few miles across which rotate thirty times a second. There are tiny grains between the stars, with the size and atomic composition of bacteria. There are stars leaving the Milky Way, and immense gas clouds falling into it. There are turbulent plasmas writhing with X and gamma rays and mighty stellar explosions. There are, perhaps, places which are outside our Universe. The universe is vast and awesome, and for the first time we are becoming part of it."

Carl Sagan,
The Cosmic Connection

Q: You have some rather strong views about the way science is handled in the popular press. Would you like to elaborate?

CS: Indeed. Science is so exciting today that I don't think it's necessary to embellish it or distort it in order to blow the minds of the readers. It's already mind-blowing enough. Such distortion doesn't convey the real excitement of science, and worse yet it encourages habits of sloppy and uncritical thinking in young readers. My frequent experience is that there is a vast popular audience enthusiastic about science, and much more willing to delve deeply than the press or TV give them credit for.

Q: There are several very interesting predictions made by Immanuel Velikovsky that turned out to be true, and I'd like your reaction to them.

First, he predicted the existence of the Van Allen radiation belt surrounding the Earth; the enormous radiation belt around Jupiter; he predicted that Mars would be found to be cratered like the Moon; he anticipated the high temperature of Venus; he also predicted that Venus would be found to be rotating in a retrograde motion beneath its dense layer of clouds, and a series of things most of which were completely against the beliefs of the astronomers at that time. What do you make of all that proven data?

CS: I make of it that Velikovsky has made a lot of wrong predictions and a few right quotations from the scientific literature. The correct quotations have been stressed and the wrong predictions have not. The right "predictions"—almost all of them—turn out to have been made by other people before Velikovsky, some of them by people whom Velikovsky himself makes reference to in his book. For example, the idea of Venus being very hot: Rupert Wildt wrote a paper in 1940 which proposed that the carbon dioxide content of the Venus atmosphere would produce a greenhouse effect which would make it much hotter than people had thought 1940 was ten years before Velikovsky's *Worlds in Collision* came out. The credit belongs to Wildt, not to Velikovsky. And that's the situation for most of the so-called "correct predictions." Some *clever* scientist saw the correct situation earlier and Velikovsky quoted him—incidentally, not always giving proper credit.

Q: And conversely, is it not also true that throughout the history of astronomy there were hundreds of wrong opinions about the size of a star, the temperature gradient of a planet, its gravitational pull, retrograde motion or whatever, and yet isn't it true that all we are given to know are the astronomers' triumphant right guesses?

CS: The progress of science is littered with dead theories; they were maladapted. But the advantage of science is that scientists—if they are any good—are willing to reject the bad ideas in favor of the good ones; that's the way progress is made. This self-correcting aspect of science is one I'd like to see more generally applied. I'd like to see politicians willing to admit that their ideas have been wrong and now they'll adopt new ones which work better. And I'd like to see popular writers of science like Velikovsky adopt similar positions. There must be a hundred items that Velikovsky was wrong on. I'd be very interested in seeing Velikovsky write a paper about all the things he was wrong about.

Q: Then, you're saying everyone should reject *everything* Velikovsky wrote?
CS: No, I don't *at all* say that one should dismiss out-of-hand the things that Velikovsky says. It's only to be dismissed *after* you read it, not *before*. I've written a ninety-page detailed critique of *Worlds in Collision*. (Unfortunately there were scientists who dismissed it before they read it.) The idea of looking at the old legends of the Earth, believing some of them and looking for cross-correlations, and deducing some natural events from them seems to me not at all an implausible method of proceeding. But when the conclusions are at variance with facts we know much more reliably—deductions, say, from the great conservation laws of physics—then we must be skeptical about conclusions drawn from myths.

Q: If you say it *doesn't* seem implausible, then you differ from Velikovsky's chief critics, because they claimed historical records were unreliable and therefore scientifically unacceptable.
CS: I think what they really are saying is that the method's unreliable, not unacceptable. I can imagine a situation where you had a very striking legend which was independently held by many diverse civilizations that you were sure had no contact with each other, and which clearly pointed to an astronomical or cosmological event about which those civilizations could have had no prior knowledge. Why, I'd be absolutely prepared to accept that such an event *had* occurred—*if* I could convince myself about the prior conditions I just mentioned. In principle there's nothing wrong with going about it that way, but you have to bear in mind that it's much riskier. Societies *do* trade legends, time scales *are* out of kilter, a story *can* have other explanations than an astronomical event.

Q: Alright. What is the likelihood of a planetary imbalance of any kind where a planet could be slowed in rotation, pulled out of orbit—any of those things Velikovsky spoke of?
CS: I think it's extraordinarily unlikely at the present time in the history of the solar system. There must have been many such events four billion years ago when the solar system was still in the process of formation, when there were a lot more colliding objects around. But the situation is very different today. Velikovsky's idea that a comet braked the Earth's rotation to a halt, and that the Earth later, somehow, started up again with the same length of the day is just plain silly—and ignores the conservation of angular momentum.

It's quite clear that we understand enough about celestial mechanics to exclude some of the events in *Worlds in Collision*. Velikovsky has to invent *ad hoc* explanations to get around the celestial mechanics—nongravitational forces, magnetic forces, and so on. The details of these new forces are never worked out, but there's plenty of hand-waving.

Q: Freeman J. Dyson claims that "the time scale for industrial and technological development for societies of alien beings is likely to be very short in comparison to the time scale of stellar evolution." He says it's probable that alien societies might be millions of years old, with science and technology of an unimaginably superior level. Their cultures, he says, will have been expanded to the limits of Malthusian principles. Suppose we suddenly made contact. Wouldn't the very existence of such a vastly superior society—even without aggressive intent on their part—be a profound psychological shock to humanity?

CS: I'm not so sure about that. The general kind of answer I'd give is that (a) the spaces between the stars are enormous, so it's not trivial for them to get here; (b) any civilization we're liable to make contact with is so *vastly* in advance of us that they could not possibly fear us yet; and (c) we're not likely to have anything that they need. I feel that the least of our problems is a direct negative consequence of receiving a message from another civilization. If we *do* make radio contact, I don't think we're going to be flooded with serious social disruptions. The existence of the message will be its most important property. We will know there is someone else out there. We will know that it is possible to survive our current period of dangerous technological adolescence—because someone else did. To understand the content of the message, to implement it, is going to be *very* slow, cautious work, taking decades or centuries.

Q: But there would have to be some kind of profound reaction.
CS: It's going to be a novelty that people will adjust to quite rapidly—assuming of course, that we're talking about a signal that takes centuries to get from there to here. I don't think it will have any important negative effects. I *do* think it will have many important *positive* effects—in drawing for us a lesson on where we are in the cosmos, and in pointing out that there may be many beings elsewhere in the universe, but only *one place* where there are human beings . . . in stressing that the organisms on this planet are

all—in the truest sense—brothers and sisters. I think our perception of *ourselves* is the principal positive consequence of contact.

Q: You've touched on the fact here that we're extremely limited by these huge gaps of time between the transmission and reception of signals. It's regarded as a kind of impassable barrier—the same sort of attitude that existed before the sound barrier was broken. Today's final, "ultimate" barrier is the speed of light; theoretically nothing can exceed that velocity. If we can't account for some of the actions of pulsars, or understand all the characteristics of quasars, doesn't this indicate that speeds exceeding that of light itself may be possible?

CS: No. I don't know of any observations of pulsars or quasars that challenge the precept of special relativity which says you can't travel faster than light. The sound barrier was never a barrier in the sense of the fundamentals of physics. It was always an *engineering* barrier. Some people thought it an insuperable engineering barrier, but it wasn't tied to the very fundamentals of physics. The idea of the velocity of light being a barrier is at the very heart of our present understanding of physics. There is a range of very strange phenomena which are repeatedly verified quantitatively—things like time dilation of very rapidly moving mesons (subatomic particles). The faster they go the slower they decay . . . the slower their little internal clocks tick. The mass of an elementary particle *increases* as it goes faster and faster, closer to the speed of light. This is why synchrotrons work.

Q: Does our current understanding presume that the theory of special relativity *can't* be wrong?

CS: The job of the physicist is to understand the way the world is put together—to make a theory which explains all the bizarre phenomena. It is one of Einstein's great achievements—not only that he was able to *explain* these things but to predict them quantitatively before they were observed, a much more difficult feat. And he did it by making some *assumptions*. One of the *fundamental* assumptions was that no material object can travel faster than the speed of light. It's an *assumption*, and being only an assumption, nothing says it can't be wrong. But that assumption permits us to understand a range of phenomena in the real world, which otherwise no one can understand at all.

Q: How do you know that tomorrow some bright fellow won't come up with a theory which will quantitatively explain all these phenomena?
CS: I don't know that there's no smart fellow who won't come along tomorrow and make such a theory. But until he *does*, I'm stuck with special relativity, which is one of the most productive and brilliant intellectual achievements of man.

Q: In what respect?
CS: In that it permits us to understand *very* strange phenomena in a very simple way, and that it's derived from a deep and simple analysis of concepts of space, time and simultaneity. It's in *that* sense that the physicist says he thinks it's true, but *only* in that sense. Because I see that I can understand many mysterious things if I believe that you *can't* travel faster than light, I believe you can't travel faster than light. But I'm quite prepared to change my mind tomorrow—if somebody comes up with a better theory. But it's *not* tomorrow. It's today. I consider special relativity very strongly supported—as strong as anything else in physics.

Q: You just spoke of time dilation. What is it, and what are its consequences?
CS: Time dilation is another consequence of special relativity which partially helps to undo the sting of not being able to travel faster than light—travel close enough to the speed of light and your local clock *can* go as slow as you want. You can get from here to anywhere else in any time you choose, provided you can go close enough to the speed of light.

Q: I don't understand that. Even so, the galaxy is about one hundred thousand light-years across, therefore even traveling at exactly the speed of light, it seems to me it would take one hundred thousand years to travel from one end of the galaxy to the other.
CS: Not at all. That's only as measured on the launch planet—or the planet to be visited. But as measured on the *spacecraft*, you could travel from here to the other side of the galaxy in—whatever you like—a year? It can easily be done. You tell me how long you want to take to get from here to some other place in the universe, and I'll tell you how fast you have to go—how *close* to the speed of light. You'll never pass the speed of light, you won't break any laws of special relativity, and still you can go *anywhere* in as short a period of time as you like. It's just an *engineering* problem to get that close to the speed

of light. It might be .99999 the speed of light, and of course there are *huge* engineering problems to ever build a spacecraft which can go that fast. But in terms of the principles of physics you can go from any point A to any point B in however short a period of time you like.

Q: Alright. How do you manage to keep up with an active membership in so many organizations and still find time for your other interests—ping pong, stamp collecting, scuba diving and such?
CS: There's something about the self-contained aspect of scuba diving that I like very much. I swim down with my camera and chase indigenous life forms—without hurting anybody. There's the sense of another biology down there, which is probably connected with my interest in finding life elsewhere in the universe. You also get a sense of three dimensions. We're very *two-dimensional* beasts walking along the surface of the Earth. In snorkeling, you know, it's very, very exciting to have twenty or thirty feet vertically within your command. With scuba gear you have a hundred or two hundred feet vertically. I'm sure that people who like skiing or gliding or sky diving do it for very much the same reasons—the thrill of that third dimension. But, I like to do it where there are other life forms around.

Q: At a meeting of the Committee on Space Research of the International Council of Scientific Unions, did any political or ideological barriers arise during the exchanges of ideas?
CS: Sure, but I'd say almost equally on both sides. But what impresses me the most is how very similar and human scientists of various nations are. The advantages of free scientific communication are enormous.

Q: How does politics intrude?
CS: For example, in the past, the Soviet Union would land an unmanned vehicle on Venus, say, or Mars. It sends along an embossed metal reproduction of the great seal of the Union of Soviet Socialist Republics. The United States sends some guys to the Moon and they plant a plastic American flag or a plate signed by Richard Nixon. Those seem to be precisely parallel activities. Scientists from both countries can deplore the intrusion of nationalism on what ought to be an international activity. It ought to be humankind that's sending unmanned spacecraft to Venus and Mars and men to the Moon.

Q: Isn't that in fact what's actually happening?
CS: In the longest perspective it is. But it would be nice to see the immense historical importance of planetary exploration acknowledged specifically. It also gives a perspective on earth-bound sciences, which is of enormous practical value. A better understanding of these matters might increase support for space science and exploration. In America today we are simply not utilizing our remarkable capability for space flight.

Q: Let's discuss UFOs; you throw out a list of alternative explanations for them—"Why the hell don't you consider this or that?"—One of the things you cited as a perfectly feasible consideration (if you're going to consider *all* the alternatives) was a time machine!
CS: I wouldn't say perfectly feasible, I would say "not obviously less feasible."

Q: Alright . . . however remote time travel may be, has anything been learned recently about the properties of time to indicate that time travel can be a natural function of the universe?
CS: No, I don't know of any new developments, along these lines.

Q: There's a paper put out by the Commerce Department called, I think, "Possibility of Experimental Study of the Properties of Time." (Ed. notes: JPRS; 45238 Published by the National Technical Information Service, Springfield, Va. 22151—$3.00) It said there were particles that supposedly move backward in time.
CS: An old idea.

Q: Theoretical or mathematical?
CS: Both. A particle moving forward in time is in some sense the same as its *anti*-particle moving *backwards* in time. Richard Feynman proposed this idea about thirty years ago. Like much of the world of elementary particle physics, it has some very surrealistic aspects. It also doesn't seem to have any practical consequences, but it's an interesting idea—another way of looking at the world.

Q: So you don't visualize anybody building time machines in the near future?
CS: No, certainly no . . . at least not anybody I know.

Q: It's been said that Edmund Halley has the most stupendous monument of any human since the dawn of history—the comet named after him when he predicted its return. Yet you were responsible for something that will probably outlast Halley's comet and everything else ever built on Earth.
CS: Halley was a terrific fellow. The *Pioneer 10* plaque has a lifetime in the depths of interstellar space of at least hundreds of millions of years. It'll be around when a lot of other things on the Earth, like the Rocky Mountains, won't be. That's because the rates of erosion in interstellar space are much slower than the rates of erosion on the surface of the Earth.

Q: The United States won't have a Mars soft-landing until the Viking mission. What did *we* learn from the Soviet *Mars-3* probe before they requested a moratorium on the news released from the data they provided? They had a hot line, and the story was that NASA could not release certain information.
CS: They wanted to release their communiqués, and we were free to release ours. They didn't want to have us release their communiqués. The *Mars 2, 3, 4* and *6* entry probes all failed, so there was very little information from them.

Q: Their entry probes failed? I understand *Mars 3* was transmitting for twenty seconds.
CS: *Mars 3*, twenty seconds of *blank* television picture.

Q: *Was* it blank?
CS: Absolutely featureless. You see, we know that at the place and time at which the spacecraft landed there were global dust storms, fierce winds . . . it was not a good place to go down. The idea that *Mars 3* failed in those high winds . . . the idea that their twenty-second television transmission was clouded out, is a perfectly plausible explanation.

Q: In the study of the red shift from the light spectra of galaxies speeding away from us, it is possible that—*relatively*—a galaxy can be moving away from us and we from it, both at *more* than half the speed of light. Can the speed at which we're flying apart *overcome* the ability of light from each galaxy to reach the other?
CS: What you're talking about is called the Law of Velocity Addition in special relativity.

Q: Well, I can certainly get around this limit on the speed of light. Suppose we have two spaceships leaving in opposite directions from the same spot. Both are going at .6 the speed of light, then relative to each other, aren't they going at 1.2 the speed of light, and therefore going faster than light?
CS: The answer to that is "no."

Q: Why not?
CS: Because that's not the way the universe works. You don't just add up the velocities. There's a *new* law when you're traveling close to the speed of light, and it's a slightly more complex equation. That complex equation *never* lets you have a relative velocity greater than the speed of light, even though the two components may be traveling as close to the speed of light as you want. Even if both are traveling at .99 the speed of light going away from each other, their relative velocity, while it's greater than .99 the speed of light, is *never* greater than 1.0 the speed of light—*you can never exceed the speed of light.*

Q: Catch twenty-two: it sounds as if it borders on the mystical.
CS: That's only because you're not in the habit of traveling at the speed of light—or close to it. It's because you're used to traveling at, say, ten miles an hour, so you sample the universe in that velocity range. If you were sampling the universe at a velocity range close to the speed of light, then what I've just said would be quite plausible to you. We must be careful not to assume that our limited personal experience applies to very different physical circumstances.

Q: It's still catch twenty-two. Doesn't it sound absurd that your clock runs slower if you run down the hill rather than stand still?
CS: That's surely not in your experience because the effect is too small to measure at ten miles an hour, and yet it's true.

Q: *Measurably* true?
CS: Yes, measurably true. But of course the faster you go, the easier it is to measure.

Q: Has this actually been done?
CS: Physicists do it all the time. Rapidly moving clocks slow down by the *precise amount* that special relativity predicts. The difference between

being right qualitatively and being right quantitatively is quite an impressive difference.

Q: These measurements must be extraordinarily small.
CS: But very accurately done. You see, you can't slip out of this one by saying, "well, it's hard to measure something that carefully." It's been measured to much finer precision than is necessary to show that it's true. And the fact that you can find mu mesons at sea level, well. . . .

Q: What are "mu mesons"?
CS: They're subatomic particles produced by the interaction of primary cosmic rays at the top of the atmosphere, and take a certain amount of time to get to sea level. *Without* special relativity the time it would take for them to get to sea level would be longer than the time it would take for them to decay into their daughter products. But *because* of special relativity, because their velocity is so close to the speed of light, their "clocks" slow down. Muons "think" it took less time to make it from the top of the atmosphere to sea level, than do observers not traveling close to the speed of light.

Q: Is the theory of special relativity applicable to widely differing life forms? Take for instance the average age of one man compared to the age of all mankind to a geologic or cosmic epoch? Doesn't a microbe with a life span that's measured in days or hours subjectively experience the same (subjective) longevity of a human being?
CS: I don't think there's any connection. All those organisms are traveling at the same speed. No one is traveling close to the speed of light. For any of the effects I've been talking about to work you have to travel close to the speed of light. The answer has to be "no." If mosquitoes always traveled close to the speed of light, then what you suggest might be the case, but they can't, at least none of the mosquitoes I know.

Q: Well, this has been enormously interesting, stimulating and enlightening. We've covered a great deal of territory—I certainly appreciate the time you've given us.
CS: I've enjoyed talking to you.

Second View: Sagan on *Encounters*
Art Harris / 1977

From the *Washington Post*, December 16, 1977. © 1977 The Washington Post. Reprinted with permission.

Carl Sagan, the forty three year-old glamor-boy of astronomy, is hunched down in the fifth row of the Ziegfeld Theater on West 54th Street, waiting for the five o'clock matinee to roll. Whooooosh! Suddenly viewers are bathed in yellow haze as a sandstorm rages across the screen.

It's Sagan's first encounter with Steven Spielberg's $20 million cosmic gamble, *Close Encounters of the Third Kind*. He watches UFO sleuth Claude Lacombe (François Truffaut) emerge from the swirling Mexican desert. Lacombe is about to fall upon a squadron of Navy fighters mysteriously lost over Florida in 1945. Where are the pilots, everyone wants to know.

Sagan scoffs. "There's not a smidgen of evidence to suggest that lights in the sky or the disappearance of ships or planes are due to extraterrestrial intervention. The return of those planes is a favorite incident of the most uncritical panderers of the Bermuda Triangle mysteries. Extraordinary claims require extraordinary evidence."

So it goes. All around, the audience sits gaping. Kids cease crunching popcorn. You can hear parents' heavy breathing, their children's gasps.

What Spielberg serves up is a tale of UFOs' visiting Muncie, Indiana, and one of the witnesses—power-company lineman Roy Neary (Richard Dreyfuss)—becoming obsessed with finding an explanation. Visiting spacecraft zip and zap about with the help of animation and Doug Trumbull's special-effects wizardry.

Carl Sagan yawns. "I'm not able to say such a thing is impossible. It would certainly be a much more interesting world if such a thing had happened. But in the real world, it hasn't.

"I find science so much more fascinating than science fiction. It also has the advantage of being true. My general feeling is that there is a lot of life throughout the universe. But feelings don't count for all that much in science. The essential point is to find out. But this movie doesn't even represent a plausible scenario."

As you might guess from the title, there's a close encounter with cosmic munchkins. This was especially displeasing to Sagan, who calls Hollywood's propensity to represent alien beings as humanoids "earth chauvinism."

"That's the part of science fiction that's so impoverished," he says. "They take a human being and warp him slightly and you get laughable caricatures of human beings. The range of possibilities is so much greater."

Carol Sagan is an exobiologist—a scientist involved in the search for extraterrestrial life. Yet, since no evidence exists to prove such a notion, he operates in that half-world between laboratory and cosmic fancy. To bridge the gap between sci-fi and hard fact, he tries to spark curiosity in the possibility of life Out There. He writes best-sellers such as *The Cosmic Connection*; at Cornell, where he runs the Laboratory for Planetary Studies, his lectures are standing room-only. Carl Sagan Productions, media incarnation, is currently preparing a thirteen-part series, *Man and the Cosmos*, for 1980 broadcasting over PBS; he continues to serve as a NASA adviser, as he did on the Mariner and Viking surveys of Mars; and he frequently tromps to Washington to lobby for giant radio telescopes capable of eavesdropping on other galaxies. Sagan feels they might be bombarding us with messages at this very moment.

Although no signal has yet been received, the scientific superstar doesn't rule out the grand possibility of life elsewhere. "I would be astounded if life weren't coursing through the cosmos," he says.

If man ever greets life in the galaxy, our neighbors are more likely to look like the Blob, or a Thing, than Barbarella, Darth Vader or Spielberg's little wobblies. The same ingredients that, eons ago, went into the cosmic blender and produced Earth are believed to be salted throughout the universe. But chances are slim, Sagan says, that the elements would ever commingle again to produce anything resembling man.

Human beings came about as sort of a "cosmic accident," he explains, though the concept might not sit well with large egos. A cosmic ray zapped a gene four and a half million years back; the gene mutated and evolved and voila, man and woman. "Most biologists would agree that if you started the Earth over again and merely let the same random factors operate, you would never wind up with anything looking like a human being. If that's the case, then some very different physical planet would have zero chance of producing the aliens you see in the movies."

It's dark and bitter cold outside as *Close Encounters* blasts off the screen and Carl Sagan, poker-faced, scouts out an escape hatch. His gray-green eyes dart down alley. No *National Enquirer* reporters lurking about.

The *Enquirer* sustains a hefty four million weekly circulation by packing its pages full of Hollywood romance, instant happiness psychic phenomena and fantastic tales of UFOs. Especially UFOs. Reporters are paid handsomely to secure quotes from "experts" that lend credence to the latest speculation, and there's a reported million-dollar bounty for absolute proof of a visitor from space. Sagan has stiff-armed the tabloid for years, but its agents keep landing on his doorsteps. A UFO endorsement from Carl Sagan would be tantamount to converting Larry Flynt.

So far, there is no evidence to support even an encounter of the first kind—sighting a UFO. Sagan calls contentions of astronauts from other worlds popping down "fundamentally silly." Yet his own speculations on extraterrestrial life draw fire from scientific peers who call *him* too freewheeling.

Such comments he repeats as a frequent guest on *The Johnny Carson Show*, where his boyish face is beamed into bedrooms across the land. People recognize him—especially people who snap up *The National Enquirer* in supermarkets. This also makes Carl Sagan "an expert."

Down the street, over an encounter with a dish of manicotti, Sagan struggles to choose between *Close Encounters* and *Star Wars*. He confesses affection for Spielberg's notion of communicating with aliens in ways other than words. In *Close Encounters*, John Williams's thematic score booms through a kind of cosmic calliope to cement a galactic friendship.

Sagan likes the idea of music as the language of the universe, and for NASA's *Voyager* flight beyond our solar system, he concocted an LP of "earth sounds" to ride shotgun. Along with "Hi there!" in sixty tongues, there is whalespeak, volcano grumble, wave crashing, animal talk—all in "evolutionary sequence"—and music. Classical, Eastern and, yes, even rock 'n' roll. If Voyager finds any space rockers beyond Pluto, they may thrill to Chuck Berry twanging "Johnny Be Good."

"There are lots of ways to communicate what we know, but few ways to communicate what we feel," says Sagan. "Music is one way to communicate emotions."

Still, he sniffs, both films, remain riddled with humanoid aliens and "scientific inaccuracies." He just doesn't understand why producers don't hire some starving grad student to guard against errors. In *Star Wars,* pilot Han Solo shoots into hyperspace in so many "parsecs." A parsec is a measure of distance, not speed. "That's like saying, 'I got up at thirty two miles this morning,'" says Sagan. And he found the award ceremony a subtle case of discrimination: characters in white receive medals for valor, while the Wookie, a minority player who braved equal hardship, was ignored.

Asked to choose between the two films, though, Sagan gives the nod to Lucas. "The eleven-year-old in me liked *Star Wars*." He banishes *Close Encounters* to the Siberia of "pop theology."

But *2001 Space Odyssey* remains his favorite sci-fi movie. It's also, of course, the only one of the three he was asked to consult on. Sagan advised director Stanley Kubrick not to depict alien beings as some nephew of the Purple People-Eater, and Kubrick's close encounter was left to the imagination. What especially worries Sagan is the negative impact such films might have on the future of space exploration. What if viewers come away from *Close Encounters* believing there is little need for a space program because UFOs will surely zip down some day, as they do in the film? Some, he fears, might even come away thinking scientists are holding back evidence as good as Spielberg's fantasies.

On the other hand, he says, "These are excellent movies for eight-year-olds. They excite a sense of wonder not too taxing for a child's mind. They may even end up intriguing children with the idea of space, and play the same role the Mars novels of Edgar Rice Burroughs played for me. If that happens and twenty years from now we have a host of young scientists who were turned on by *Star Wars* and *Close Encounters*, Hollywood will have performed a service. But we'll have to wait and see."

Carl Sagan's *Cosmic Connection* and Extraterrestrial Life-Wish
Dennis Meredith / 1979

From *Science Digest*, June 1979, pp. 34+. Copyright © 1979 Dennis Meredith. Reprinted by permission.

Next month, a gangling one thousand and eight hundred pound robot named *Voyager 2* will whisk past the planet Jupiter at fifty eight thousand miles per hour, transmitting back to earth-bound scientists more detailed color pictures of an immense globe of colorful icy gases.

The scientists, astounded by the pictures returned by *Voyager 1* in March, and already puzzled enough by the intricate, swirling patterns of Jupiter's atmosphere, were mystified by the first closeups of four of Jupiter's moons—exotic, tiny worlds of ice and rock and sulfur and salt.

In September, a more modest probe, *Pioneer 11*, will become the first man-made object to encounter the elegant ringed planet Saturn.

All these machines are headed beyond our solar system, into interstellar space. Besides their sophisticated cameras and sensors, the probes carry messages meant to tell any aliens who might discover the derelict spacecraft in some far distant future about the peculiar creatures who built and launched them.

The Pioneer message is a simple gold-plated plaque etched with basic data about humans and their solar system, and the Voyager missive is an elaborate record, on which is encoded a portfolio of the sights and sounds of Earth.

Both messages arose from the efforts of the same man: Carl Sagan, the seeker of extraterrestrial life whose quest has taken him from the laboratory bench to the TV talk show couch. The messages reveal perhaps as much about the man who inspired them as they do about mankind in general. Like this noted astronomer and Cornell professor, they follow a strict academic logic. And like the Sagan who is author of the popular books *The Cosmic Connection* and *The Dragons of Eden*, the spacecraft messages are lyrical and aesthetic. Both the messages and the sender emit a personable glow—one because of its gold patina, the other because of dark good looks, a ready wit, and a casual penchant for turtle-necked sweaters. No doubt, if we have anything at all worth saying to multi-eyed, purple-tinged creatures ten million years from now, Dr. Sagan has said it attractively for us.

But also like Carl Sagan, those space-borne introductions to humanity demonstrate some effective salesmanship aimed at the folks back home. Even though those "cosmic messages-in-a-bottle" are more like grains of sand on a galactic beach—extremely unlikely to be found—they were sent anyway. And with that act, Dr. Sagan reminded us that other creatures *are* probably out there on other worlds to explore . . . and his messages are invitations to us to do just that.

Dr. Sagan's academic specialty is planetary chemistry and physics, but he also has been deeply involved in the search for alien life. This field known as exobiology is sometimes called a science with no data. No hard evidence ever has been found that life exists on other planets. Nevertheless, the great majority of astronomers are almost certain that all sorts of galactic beasties do populate the universe. For one thing, so many stars are in the universe—100 billion at last count in our galaxy alone—that statistics almost dictate that life does exist elsewhere. And radio astronomers have detected emanations from interstellar dust clouds that reveal the basic chemical building blocks of life—carbon compounds, ammonia, etc.—throughout the galaxy.

So, even though the jury is out (in fact, light-years out) the verdict probably will be favorable eventually, and Sagan has involved himself in numerous projects to gather further evidence. For example, he was a member of the biology team that searched for life on Mars via the Viking landers (with ambiguous results) and has been involved in thus far unsuccessful efforts to detect radio signals that might be beamed at earth from advanced civilizations.

Critics charge that Dr. Sagan has an "extraterrestrial life-wish," a compulsion to look for life where the chances are slim or nonexistent.

"I believe that the search for life is of such extreme importance to science, philosophy, and to our ideas about ourselves that every time we go to a new place, we have to ask ourselves seriously about whether there's life there," he responds. "But sometimes people confuse the serious asking of the question with some prior commitment to an answer."

So, Carl Sagan is willing to look for life, not only where it might be, as on Mars, but even where it's probably not, as in the frigid, rolling clouds of

Jupiter, or on the mysterious surface of Titan, the Saturnian moon which is the only one known to have an atmosphere, perhaps of organic compounds.

"I don't think there *has* to be life on Jupiter," he says, "but I can imagine that there *might* be. It's a thing worth checking, as it is for Titan. If it turns out that the solar system is lifeless except for Earth, that's an important statistic, and if it turns out there's life elsewhere in the solar system, *that's* an important statistic."

Thus, Dr. Sagan is currently a member of the Voyager imaging team, along with his colleagues poring over the pictures and other data to sort out the kinds of chemical compounds which produce the colors of Jupiter, and later of Saturn.

What kinds of life might be slithering, crawling, rolling, or flying across the surfaces of alien planets?

Like his colleagues in the field, Carl Sagan believes that alien life is probably carbon-based, built from compounds similar to those in our own bodies.

"I don't like to be a carbon chauvinist," he says, "but I keep finding that the physics and chemistry force me in that direction."

Silicon, the other major candidate as a basis for life, just cannot form the wide range of nonrepetitive, information-coding molecular chains such as DNA, Dr. Sagan asserts.

"The only circumstances in which silicon might be able to form nonrepetitive molecules, in which the units could carry genetic information, is if they formed compounds such as polysiloxanes. These kinds of compounds are likely only in environments in which the temperatures are very low, and therefore in which carbon compounds would be in even greater abundance.

"The places where people customarily think about silicon-based life—hot planets, because there are silicates there—miss the point that silicate molecules are only mindless repetitions of the same unit, so I can see no possibility of their containing genetic information."

All this is not to say that the average alien would resemble us, despite the tendency of TV and movie aliens to look like nothing more than unimaginative humans at Halloween.

"Organisms on earth are a product of the stochastic aspect of the evolutionary process. Start the Earth out over again, and let random factors operate, and you're not likely to come up with anything like a human being. Therefore, if you have a quite different physical environment, and if there's life, it will be phenomenally different," Dr. Sagan speculates.

His leaps of speculation and theory building are what has drawn the principal critical fire from some colleagues. They accuse him of grandstanding, of setting up the public for disappointment if sensational theories don't pan out. Perhaps some truth resides in such criticism, and perhaps also some jealousy exists on the part of less articulate or less sought-after colleagues.

But the major Sagan no-no, it appears, has been *to do science out loud.* His public speculation is really no more far-out than that done by most scientists in private. The difference is that he bucks the scientific tradition of keeping half-baked theories safely locked away in one's head until they can be either proved or discarded.

As a confirmed science junkie, Dr. Sagan is intent on turning the public on to the delights of science, and so he enjoys showing how it's done. His books seek not only to explain science but to link it with other human endeavor. They are replete with poetry, phrases from history's great thinkers, and works of art.

"If there's one thing that I want for the enterprise of science—astronomy in particular—I want it to be seen as a human endeavor, as a characteristic thing that people do," he explains. "No other species on this planet 'does' science. Other species have strong emotions; it's not our emotions that make us unique. It's our way of thinking, and I think science exemplifies this best.

"I'd like to believe that people are designed to enjoy that kind of thing, but that society is arranged to discourage them in the early school years.

"I believe that in every person is a kind of circuit which resonates to intellectual discovery—and the idea is to make that resonance work."

Currently, Dr. Sagan is working on an ambitious television series that he hopes will make our intellectual circuits resonate furiously. The thirteen-part, $8 million science series, called *Cosmos*, will be televised next spring on the Public Broadcasting System. It might well be called *The Cosmic Correction,* for Dr. Sagan hopes the series, combining dazzling special effects with scrupulous scientific accuracy, will go a long way toward correcting the abysmal state of science in the electronic media.

No doubt, a good science series can help. Science news reporting on television is perfunctory and boring, and so-called science fiction in TV and movies is nothing more than cowboys and aliens, with poor science caught in a withering crossfire of death rays and photon torpedoes. And in TV and movies, and in books as well, the public is continually suckered by fairy tales of ancient astronauts, Bermuda triangles, listening plants, and lined-up planets producing catastrophic earthquakes.

"In *Cosmos* we'll have location shooting, exquisite special effects, and in the studio a 'Spaceship of the Mind' that will take us wherever we want to go, in both space and time," Dr. Sagan predicts. "I'd like the series to be so visually stimulating that somebody who isn't even interested in the concepts will watch just for the effects. And I'd like people who are prepared to do some thinking to be *really* stimulated."

If Dr. Sagan has his way, "stimulating" will be a mild term for *Cosmos*. Plans for the series include a wild, looping ride through the universe, an excursion into the body and through a living cell, a tour of the intellectual treasures of the Great Library at Alexandria, and strolls on the planets.

Although he will use astronomy as a starting point for *Cosmos*, Dr. Sagan plans to journey widely in the realm of human experience. Even before it has begun, however, his prospective voyage already is being met by complaints from some fellow scientists that this time he is venturing too far outside his field.

"We're concentrating on how astronomy—the cosmos—relates to human beings, in terms of how the atoms inside our bodies were constructed inside stars, in terms of the history of life on Earth having been determined by cosmic events, and how our philosophies and myths are in many ways tied to astronomical themes," is the Sagan rebuttal.

"Another theme will be this remarkable branch point in the history of our species that we're at now—for the first time stepping off the Earth a little bit and looking around. We'll go into the history of exploration on the Earth and make analogies between sailing ship exploration and space ship exploration."

Cosmos will be not only an intellectual statement, however, but a political one, for Dr. Sagan hopes the series will generate more public enthusiasm for astronomy and space exploration. Despite his advocacy of voyages to other worlds, though, he opts against a large-scale Apollo-type program.

"Scientists constantly get clobbered with the idea that we spent twenty-seven billion dollars on the Apollo programs, and are asked 'What more do you want?' We didn't spend it; it was done for political reasons," he declares, explaining,

"Apollo was a response to the Bay of Pigs fiasco and to the successful orbital flight of Yuri Gagarin. President Kennedy's objective was not to find out the origin of the moon by the end of the decade; rather, it was to put a man on the moon and bring him back, and we did that."

Thus, although Carl Sagan does believe that technological advance eventually will give humans a permanent niche in outer space, he advocates a less expensive program of unmanned exploration for the immediate future:

"We have to ask what space program will have the same hopeful, forward-looking benign implications (as the Apollo program), but will not cost hundreds of billions of dollars. Manned missions to the planets? Large colonies in earth orbit? Manned bases on the moon? Large-scale solar-power stations in earth orbit? All these cost hundreds of billions of dollars.

"What costs a lot less? Three things: a vigorous program of unmanned exploration of the solar system; a search for planetary systems around other stars using earth satellites and orbiting telescopes; and a radio search for extraterrestrial intelligence. A vigorous pursuit of all three would cost 1 percent of these other things, and therefore we could afford it."

Dr. Sagan and his colleagues have their work cut out for them. Despite the spectacular successes of Voyager and other planetary missions and the forthcoming debut of the Space Shuttle, the country is in a period of exploratory timidity.

And it seems not to be a time to advocate hurling robots into space so that we can become vicarious cosmic tourists.

But perhaps it should be. Considering our current low national morale, what Carl Sagan is selling—a little pride, a little adventure, a few dreams, and some knowledge—might prove to be worth the price.

The Cosmos
Jonathan Cott / 1980

From *Rolling Stone*, December 25, 1980–January 8, 1981. © 1981 Rolling Stone. All rights reserved. Reprinted by permission.

"We are a way for the cosmos to know itself," Carl Sagan stated on *Cosmos*, his recently broadcast thirteen-part series on public television. As he recreated journeys back in time and through the universe, speculating on its future and ours, Sagan continually reminded us that fresh knowledge of reality, even that which signals change, is inspirational, not dangerous.

In his essay "In Praise of Science and Technology," Sagan writes: "The most effective agents to communicate science to the public are television, motion pictures and newspapers—where the science offerings are often dreary, inaccurate, ponderous, grossly caricatured or (as with much Saturday-morning commercial television programming for children) hostile to science." Sagan has attempted to correct this balance in his best-selling books and frequent appearances on television talk shows, but *Cosmos* has been his most ambitious and sustained undertaking to date. In the series, he used extraordinary special effects and a remarkably uncondescending, popular approach to present scientific information, displaying what one poet defines as the Homeric style: "eminently rapid, plain, direct in thought, expression, syntax, words, matter, ideas, and eminently noble." This proved to be an eminently suitable style with which to communicate deep and fundamental ideas about the universe to the close to 150 million people around the world who viewed the series.

In addition to his television work, Sagan is director of the Laboratory for Planetary Studies and is the David Duncan professor of astronomy and space sciences at Cornell University, where he also serves as associate director of the Center for Radio-physics and Space Research. He played a leading role in the Mariner, Viking and Voyager expeditions, and he is the author of such books as *The Cosmic Connection; The Dragons of Eden*, for which he won a Pulitzer Prize; *Broca's Brain*; and *Cosmos*, which is based on the series.

Joining in the following interview is Ann Druyan, who, along with Steven Soter, contributed to the *Cosmos* scripts. The conversation took place at Sagan's Los Angeles home in late August while he put the final touches on *Cosmos*.

Jonathan Cott: In your book *The Cosmic Connection*, you quote T. S. Eliot: "We shall not cease from exploration/And the end of all our exploring/Will be to arrive where we started/And know the place for the first time." I want to focus on the word "know" and ask you about knowing things for the first time, since this seems to be a seminal notion in your work.

Carl Sagan: We start out a million years ago in a small community on some grassy plain; we hunt animals, have children and develop a rich social, sexual and intellectual life, but we know almost nothing about our surroundings. Yet we hunger to understand, so we invent myths about how we imagine the world is constructed—and they're, of course, based upon what we know, which is ourselves and other animals. So we make up stories about how the world was hatched from a cosmic egg, or created after the mating of cosmic deities or by some fiat of a powerful being. But we're not fully satisfied with those stories, so we keep broadening the horizon of our myths; and then we discover that there's a totally different way in which the world is constructed and things originate.

Today, we're still loaded down, and to some extent embarrassed, by ancient myths, but we respect them as part of the same impulse that has led to the modern, scientific kind of myth. But we now have the opportunity to discover, for the first time, the way the universe is in *fact* constructed, as opposed to how we would wish it to be constructed. It's a critical moment in the history of the world.

Jonathan Cott: The Eliot quote also seems to suggest that, as explorers, human beings may exist to explain the universe to itself.

Carl Sagan: Absolutely. We are the representatives of the cosmos; we are an example of what hydrogen atoms can do, given fifteen billion years of cosmic evolution. And we resonate to these questions. We start with the origin of every human being, and then the origin of our community, our nation, the human species, who our ancestors were and then the riddle of the origin of life. And the questions: where did the earth and solar system come from? Where did the galaxies come from? Every one of those questions is deep and significant. They are the subject of folklore, myth, superstition and religion in every human culture. But for the first time we are on the verge of answering many of them. I don't mean to suggest that we have the final answers; we are bathing in mystery and confusion on many subjects, and I think that will always be our destiny. The universe will always be much richer than our ability to understand.

For example, Io, one of Jupiter's big moons, was undiscovered until the seventeenth century. Until 1979, it was a point of light in the view of all but the few astronomers who had access to very large telescopes and could see the faintest mottling on the surface. Now we have thousands of detailed photographs showing features a kilometer across. We have passed from ignorance to knowledge of a whole world. Well, that's just one world. There are twenty other planets and moons we have since photographed. Twenty new worlds.

Jonathan Cott: Freud wrote about the moment when an infant sees himself in the mirror for the first time.
Carl Sagan: That's a very good metaphor; we've just invented the mirror, and we can see ourselves from afar.

Jonathan Cott: In the *Cosmos* series, you stated that the fact that the universe was knowable was attested to in the sixth century BC in Greece.
Carl Sagan: Sixth-century Ionia was, to the best of my knowledge, the first time there was a generally accepted view that the universe was subject not to the whims and vagaries of the gods but to generally applicable laws of nature that human beings were able to understand.

It wasn't until the 1960s that the first photograph of the whole earth was taken, and you saw it for the first time as a tiny blue ball floating in space. You realized that there were other, similar worlds far away, of different size, different color and constitution. You got the idea that our planet was just one in a multitude. I think there are two apparently contradictory and still very powerful benefits of that cosmic perspective—the sense of our planet as one in a vast number and the sense of our planet as a place whose destiny depends on us.

Jonathan Cott: You've often quoted the Russian scientist K. E. Tsiolkovsky's statement: "The earth is the cradle of mankind, but one cannot live in the cradle forever."

Carl Sagan: I strongly dislike the notion that if things get absolutely rotten here, we can run away to somewhere else. I think it's a silly idea on economic *and* on moral grounds. Nevertheless, it's true, in my opinion, that the maturity of the human species will be connected with our ability to leave the earth, our mother, and seek our fortune in the galaxy . . . but not to abandon the earth, by any means. If we don't put our house in order, we'll never be able to explore the cosmos.

Life has had four billion years to develop through tortuous trial and error. But unlike biological evolution, which is fundamentally a random process and extremely wasteful in terms of lots of organisms dying, we don't have that opportunity. If we destroy ourselves, it may be a minor tragedy for life on the planet, but it's certainly a major tragedy for us. So we have to foresee the mistakes and avoid them. We can't stumble and then say, "I guess next time stockpiling fifteen thousand targeted nuclear warheads is not a good thing. I've learned from my mistake." I think there's a serious danger of our civilization destroying itself, and at least a possibility of our species destroying itself. But the destruction of all life on Earth is unlikely, and certainly we can't destroy the planet. There's a hierarchy of destructibility.

Jonathan Cott: Today, we can possibly destroy not only ourselves but also, it seems, some of our most intelligent hypotheses. More and more people, for example, are agreeing with Luther Sunderland, the New York spokesman for the "creationists" [antievolutionists]. Sunderland says: "A wing is a wing, a feather is a feather, an eyeball is an eyeball, a horse is a horse, and a man is a man."

Carl Sagan: The theory of evolution is the best explanation by far of the beauty and diversity of the natural world, and it's hard to see how evolution by natural selection wouldn't work. I think a fundamental problem with people who have trouble with the idea of evolution is the time perspective. You stand around, you watch a tree; it doesn't turn into anything else. You say, "This evolution stuff is nonsense." But wait a hundred million years and you will see something quite different. That instinctive feeling—"If I haven't seen it, it doesn't exist"—is, I think, behind some of the doubts on evolution. But it's also behind some of the doubts people have about special relativity. Special relativity says that if you travel close to the speed of light, your watch slows down and you can travel into the far future. Or quantum mechanics says that, in the realm of the very small, you can have a dumbbell-shaped molecule in this position or that position but not in any intermediate position. "Well, ridiculous, I never saw any rule that prevents me from turning a thing to any intermediary position I want."

This is an example of the inapplicability of common sense. Common sense works fine for the universe we're used to, for time scales of decades, for a space between a tenth of a millimeter and a few thousand kilometers, and for speeds much less than the speed of light. Once we leave those domains of human experience, there's no reason to expect the laws of nature to continue to obey our expectations, since our expectations are dependent on a limited set of experiences.

That's part of the disquiet a few people feel about evolution. Also, some people are annoyed by the idea that we are not the apex of the universe.

Jonathan Cott: They'd rather be the apex than the ape.
Carl Sagan: If I thought the supreme coordinator of the universe had a special interest in making me and my brothers and sisters, that would give me a special significance. It would make me feel good, and also make me think that maybe I didn't have to take care of myself; someone much more powerful would do so. It's a tempting idea, but we have to be very careful not to impose our hopes and desires on the cosmos, but instead, in the scientific tradition and with the most open mind possible, see what the cosmos is saying to us.

On the question of creationism, it is true that natural selection as the cause of evolution is a hypothesis. There are other possibilities. The creationists argue that they're interested in fairness: they don't want only one of several competing doctrines taught in the schools. I applaud their interest in fairness, but I think that the first test is their willingness to teach Darwinian evolution in the churches. If they're worried that there isn't fair exposure of both sides, then it's quite remarkable how only one side is taught in the churches, the synagogues, the mosques and, I might add, during the enormous number of hours on television devoted to presenting idiosyncratic belief systems.

Jonathan Cott: In your books and throughout the *Cosmos* series, you seem to be deeply committed to the idea of the relatedness and connectedness of all universal material.
Carl Sagan: It's a truth of enormous power. Talk about things that ought to be shouted from the pulpits—this is surely one. The matter we're made out

of was cooked in the center of stars. We're made of star stuff—the calcium in our teeth, the carbon in our genes, the nitrogen in our hair, the silicon in our eyeglasses. Those atoms were all made from simpler atoms in stars hundreds of light-years away and billions of years ago.

It's an astonishing thing, we're so tied to the rest of the cosmos. Cosmic rays that are produced in the death throes of stars are partly responsible for the mutations that have led to us—the changes in the genetic material. The origin of life was spurred by ultraviolet light from the sun and lightning, which in turn is caused by the heating of the earth by the sun. The connections are intricate and powerful and lovely. For those people seeking a cosmic tie-in, one exists. It's not the one the astrologers pretend, but it's much more elegant, and it has the additional virtue of being true.

Jonathan Cott: I know you're not an avid consulter of astrologers.
Carl Sagan: I'd be all for it if there were any evidence for it, but there isn't. It's like racism or sexism: you have twelve little pigeonholes, and as soon as you type someone as a member of that particular group, as long as someone is an Aquarius, Virgo or Scorpio, you know his characteristics. It saves you the effort of getting to know him individually.

Jonathan Cott: In his book *The Natural History of the Mind*, Gordon Rattray Taylor distinguishes, as you do not, between the mind and the brain, and he gives as examples things that he thinks can't be adequately explained by studying only the brain, such as altered states of consciousness, amnesia, artistic inspiration, imagination, inhibitions, pain, placebo effect, sight, smell, telepathy, willpower and love.
Carl Sagan: Talk about imagination! What a lack of imagination in the contention that those things can't be. . .
Ann Druyan: . . . based on material reality.
Carl Sagan: Right. I mean, for example, he mentions altered states of consciousness. Look how psychedelic drugs, like alcohol, regularly produce altered states of consciousness. It's a simple molecule: C_2H_5OH. Put that in your system and suddenly you're feeling very different. Well, is that mystical, or does that have something to do with chemistry?

Jonathan Cott: You're talking about what this chemistry causes rather than about what you're experiencing in that state of consciousness.

Ann Druyan: But why separate the dancer from the dance? Why separate the experience from what causes the experience? It's not necessary. The whole idea of science is to trust in reality and to interrogate nature so you can get answers, can step right up to the mirror—reality itself—and not turn away from it.

Jonathan Cott: In *The Dragons of Eden*, you write that "it is because of this immense number of functionally different configurations of the human brain that no two humans, even identical twins raised together, can ever be really very much alike. . . . All possible brain states are by no means occupied; there must be an enormous number of mental configurations that have never been entered or even glimpsed by any human being in the history of mankind." What do you think will enable human beings to occupy these configurations?
Carl Sagan: Well, I don't know. There are many that may not be entered by a single person within the next thousand years.

Jonathan Cott: What can human beings do to try to enter into these areas?
Carl Sagan: One thing to do is to mistrust the conventional perceptions. If you're interested in a new perception, you have to view with some degree of objectivity still-unspoken truths.

Jonathan Cott: So the scientists you talk about in *Cosmos* are quite subversive?
Carl Sagan: Yes. As Alfred North Whitehead said, "It is the business of the future to be dangerous." Any new idea that doesn't threaten something isn't worth its salt.

Jonathan Cott: Do you think the future is going to be dangerous?
Carl Sagan: Absolutely. The present is quite dangerous also, though. Let me give you an example. I think it's clear that none of the forms of government that exist in any of the two hundred or so countries on the earth today are applicable to the middle of the next century. Not a one. We have to get from here to there somehow. How can you do that without disturbing the *here*? The world is changing at an incredibly rapid pace. Human survival depends on dealing with those changes, but governments generally are concerned with changing nothing.

I think that any nation with a serious concern about the future would be busy inventing experimental communities to try, on a practical basis, to find the society that is going to work in the middle of the twenty-first century. I think the alternative communities of the sixties were a premonition, a spontaneous recognition by a lot of people that society, by and large, wasn't working, and that they had to see what else they could do. The larger society was unhappy with the idea of alternatives. The possibility of a better world is a rebuke. It says, "Why haven't you worked to make that change?" Since very few of us manage to make any significant changes, we tend to resist that exhortation.

Ann Druyan: There is a resistance to change, but there is no refuge from change in the cosmos. So it's a very grave problem.

Jonathan Cott: So you're trying to wake people up a bit.

Carl Sagan: Those are highly ethical motivations. But a lot of my motivation is that understanding science is fun. It's communicable fun.

Jonathan Cott: You don't want to be portentous.

Carl Sagan: Science, as communicated in some places, sounds as if it were the last thing in the world that any reasonable person would want to know about. It's portrayed as impossibly difficult to get into and a thing that sort of rots your brain for any good social interaction.

Jonathan Cott: On "Slow Train Coming," Bob Dylan refers to scientists in a very disparaging manner.

Ann Druyan: I take this very deeply to heart. The thing that I always loved about Dylan was the courage of his metaphors and the way he could cut to the bone of some kind of naked feeling. It always seemed very gutsy. And now it seems that he's turned away, he's blinded by the light, and so he looks for some easy explanation.

Jonathan Cott: In *The Dragons of Eden*, you quote St. Augustine of Hippo, who said, "I no longer dream of the stars."

Carl Sagan: Just compare that with another quote: "To dance beneath the diamond sky with one hand waving free." Compare that with Augustine and with Dylan's latest incarnation.

Jonathan Cott: Concerning your notion of the enormous amount of mental configurations in our brains, you've written: "From this perspective, each human being is truly rare and different, and the sanctity of individual human lives is a plausible, ethical consequence." This connects with another of your remarks concerning "the profound respect for other human beings and organisms as coequal recipients of this precious patrimony of 4.5 billion years of evolution." Both communicate a very Buddhist sense of the importance of the love for all sentient beings and creatures.

Carl Sagan: Don't you think that's just a logical extension? People certainly love their families, then distant relations, then friends; then they have some degree of affection for their community, their tribe. One principal level of human identification right now is with the nation-state. Now, the obvious next identification is with all the people on the planet. But why is that the end? I mean, especially if we understand our common heritage, our genetic relationship to animals and plants. Why not a set of absolutely continuous dissolves, one animal to another? Don't we have some degree of sympathy and respect for all the living things on the planet? They are our cousins. It's such an obvious idea.

Jonathan Cott: Your perspective is ethically far wider than the one we generally see operating today.

Carl Sagan: It's the time-perspective point again. Most of human history was spent in hunter-gatherer communities. And in these kinds of communities today—there aren't many of them—you find a degree of cooperativeness, an absence of alienation that is unheard of in modern society. To ignore our social heredity is a serious mistake. There is a human capacity for good-natured cooperation that is simply not encouraged in modern society. That must change.

Jonathan Cott: In the scientific world there are such subjects as particle physics, astrophysics, biophysics and geophysics—all these compartmentalized and specialized areas. People working in any one of these areas are often afraid to make general statements about matters outside their domain. Yet in *Cosmos*, you take on the entire cosmos!

Carl Sagan: It's fun to do. It's certainly where the excitement is—on the border of two fields that haven't made much contact yet. The boundaries are arbitrary. Those things that separate, say, astronomy from geology, or chemistry from biology, or even mathematics from physics, these are manmade, human-invented boundaries. In the real world, these subjects flow into each other.

Everything is related. Suppose there's a computer that goes through the names of everybody in the country and randomly picks out one person, and you have to get in touch with that person. You have to call someone, who in turn has to call someone else, and so on. What's the average number of calls you'd have to make to get that targeted person?

I mean, how many people could you call who would recognize you, even vaguely, so that you could say, "Hello, Charlie, sorry to bother you. I know you live in Omaha, but there's a guy in Fargo, North Dakota, I'm trying to get in touch with. Would you mind making one call for me?" How many people do you know who would make a phone call for you to someone he knows whom he could ask the same question? How many, just roughly?

Jonathan Cott: Maybe seventy or eighty.
Carl Sagan: Let's round it off to hundred. Let's suppose that's true of everybody. So you know hundred people, and suppose each of them knows hundred people—only a few of whom are already on your list. So to get to ten thousand people, that's just two calls—hundred times one hundred. To get to a million is three, to get to hundred million is four, and there are only two hundred million people in the country.

Jonathan Cott: So what is the moral of this example?
Carl Sagan: That it's not just some peculiar idea of the Buddhists. It's the truth: everything is connected.

Jonathan Cott: The *New York Times* reported not long ago that one bewildering outcome of quantum theory has led some scientists to speculate that the entire universe, "including the time in which it exists, may have been created by a spontaneous quantum fluctuation—a 'twitch' in the nothingness that preceded it." That sounds a lot like Buddhism, doesn't it?
Carl Sagan: I agree. That does sound like an Eastern religion. And it may be based on a perfectly respectable scientific paper.

Jonathan Cott: This kind of speculation leads to religious and philosophical questions, doesn't it?
Carl Sagan: All of science does. I think that's why we *have* religious questions: because we are naturally scientists. It's the only thing we do substantially better than other creatures. Even much of our music is an expression of feelings that we share with other animals but actualize because we're good at science and technology and they're not. And science and technology—surely no other animal on the planet has it, aside from termite nests and so on;

that's a distinctively human ability. Feelings are not characteristically human—very likely animals have lots of deep feelings. It's thinking that's characteristically human. So I don't think you should be surprised that a religious idea turns out to have some scientific support.

Jonathan Cott: But you've mentioned that science is still a myth.
Carl Sagan: Well, a myth is an attempt to pull together the best information that's available to explain the origin of something.

Jonathan Cott: So there may possibly be a better myth than science in the future?
Carl Sagan: It's guaranteed. How likely is it that we live in the very year that the absolute truth is first found out about the cosmos? It would be a remarkable coincidence, considering how many years there are. It's much more likely that human knowledge is a set of successive approximations and that there are all sorts of things that we've gotten wrong, and all sorts of mind-boggling things that we can't even *glimpse* that will be the established fact in a century or two.

Jonathan Cott: You're saying that there are ways of thinking that we know nothing about.
Carl Sagan: Must be. On many different levels the answer to that must be yes. T. S. Eliot talks about knowing a place for the first time. But there's a second and a third time. I think there's a continuum of fractional times. You always know the earth to some degree, you always know home to some degree, but you can always make significant increments in your knowledge of them.

Jonathan Cott: So there's never a certitude?
Carl Sagan: There are two extremes to worry about. One is the extreme in which everything is known and there's nothing left to do. The other is where everything is so complicated you can never begin to do anything. We are lucky to live in a universe where there *are* laws of nature and things to discover, but they're not impossibly difficult, so we can understand them to some extent. But they're also difficult enough so that we're nowhere near understanding them all. There are exhilarating discoveries yet to be made. It's the best possible world.
Ann Druyan: The best possible cosmos!

God and Carl Sagan: Is the Cosmos Big Enough for Both of Them?
Edward Wakin / 1981

Copyright 1981 *U.S. Catholic*. Reproduced by permission from the May 1981 issue of *U.S. Catholic* (pp. 19–24). Subscriptions: $22/year from 205 West Monroe, Chicago, IL 60606; call 1-800-328-6515 for subscription information or visit http://www.uscatholic.org/.

Carl Sagan has been described as America's "most effective salesman of science." A colleague at Cornell University, where Sagan is professor of astronomy and director of the Laboratory for Planetary Studies, has compared him with academic scientists: "He is very often right and always interesting. That is in contrast to most academics, who are always right and not very interesting."

He is, above all, the best-known science teacher in the country, the embodiment of science. His now-familiar face is watched by millions of viewers of *Cosmos*, the celebrated series on public television. Sagan's audience of readers as well as viewers is formidable and his books, among them *The Cosmic Connection*, *Dragons of Eden*, and *Murmurs of Earth: Voyager Interstellar System Studies*, have sold millions of copies and have been translated into a dozen languages.

Sagan insists that "there is nothing about science that cannot be explained to the layman." In his books and in his television series, this Pulitzer Prize–winning scientist, author, and teacher proves his point. No one makes science come alive so clearly as does this forty-five-year-old astronomer who is at home in TV studios, classrooms, laboratories, and even the U.S. space program.

For religious believers, he is particularly interesting. In one personable and exciting individual, the believer can encounter the thinking, the attitudes, and the views of modern science. No one elected Carl Sagan, the boy from Brooklyn who dreamed of studying the cosmos, to be spokesperson for science. He did not seek the title. But he has it and if a religious believer wants to know what scientists think of belief and believers, then there is no better witness than Carl Sagan.

Edward Wakin / 1981

EW: In a Sunday sermon you once gave in Sage Chapel at Cornell University, you commented that the confrontation of religion and science has "eroded" traditional religious views, "at least in the minds of many." What is happening between science and religion today?

CS: Broadly considered, a religious attitude and often some religious content is part of virtually every scientific investigation. If we look at the universe in the large, we find something astonishing. We find a universe that is exceptionally beautiful, intricately and subtly constructed. Whether our appreciation of the universe is because we are a part of that universe, evolved in it and by it, is a proposition to which I do not pretend to have an answer. But there is no question that the elegance of the universe is one of its most remarkable properties. It is very hard to look at the beauty, intricacy, and subtlety of nature without feeling awe. I don't think even the word reverence is too strong.

EW: Where does God fit into this view?

CS: When people ask me after one of my lectures, "Do you believe in God?" I frequently reply by asking what the questioner means by "God." The term means a lot of different things in a lot of different religions. For some, it's an outsized, light-skinned male with a long white beard, sitting on a throne somewhere up there in the sky, busily tallying the fall of every sparrow. To others—for example, Baruch, Spinoza, and Albert Einstein—God is essentially the sum total of the physical laws which describe the universe. I can't imagine anyone denying the existence of the laws of nature, but I don't know of any compelling evidence for the old man in the sky.

In the cosmic context, the very scale of the universe—more than one hundred billion galaxies, each containing more than one hundred billion stars—speaks to us of the inconsequentiality of human events. We see a universe simultaneously very beautiful and very violent. We see a universe that does not exclude a traditional Western or Eastern god, but that does not require one either.

EW: Still there is the question of a "first cause" that is of concern to religious believers, particularly in the Christian West.

CS: I would say the question of a "first cause" is only a speculation. It's perfectly possible that the universe is infinitely old and therefore uncaused. In fact, there are detailed cosmological models that hold such a view and that are consistent with everything we know. To my mind, it seems not fully satisfactory to say that there was a first cause. That seems to postpone dealing with the problem rather than solving it. If we say "God" made the universe, then surely the next question is, "Who made God?" If we say "God" was always here, why not say the universe was always here? If we say that the question "Where did God come from?" is too tough for us poor mortals to understand, then why not say that the question of, "Where did the universe come from?" is too tough for us mortals? In what way, exactly, does the God hypothesis advance our knowledge of cosmology? What predictions does it make on which the hypothesis will stand or fall?

EW: That seems to leave the question up in the air as far as you are concerned.
CS: Those who raise questions about the God hypothesis and the soul hypothesis are by no means all atheists. An atheist is someone who is certain that God does not exist, someone who has compelling evidence against the existence of God. I know of no such compelling evidence. Because God can be relegated to remote times and places and to ultimate causes, we would have to know a great deal more about the universe than we do now to be sure that no such God exists. To be certain of the existence of God and to be certain of the nonexistence of God seem to me to be the confident extremes in a subject so riddled with doubt and uncertainty as to inspire very little confidence indeed. A wide range of intermediate positions seems admissible. Considering the enormous emotional energies with which the subject is invested, a questing, courageous, and open mind is, I think, the essential tool for narrowing the range of our collective ignorance on the subject of the existence of God.

EW: Then, am I correct in finding you open on the matter of God? You appear to feel that the jury, particularly the jury of scientific experts, is still out. You seem intent on finding natural explanations for things that might be ascribed to the religious and the supernatural.
CS: Yes, except that I would say there is no deeper religious feeling than the feeling for the natural world. I wouldn't separate the world of nature from the religious instinct. Einstein, among others, made that point very strongly in his appreciation of the depth and beauty of the universe, which he described as a religious experience. To quote him: "In this sense, and in this sense only, I belong to the ranks of the devoutly religious men."

EW: How then do you feel about believers and nonbelievers?
CS: I have some discomfort both with believers and with nonbelievers when their opinions are not based on facts. I am extremely uncomfortable with dogmatic atheists, who claim there can be no God; to my knowledge, there is no strong evidence for that position. I'm also uncomfortable with dogmatic believers; to my knowledge, they don't have any strong evidence either. If we don't know the answer, why are we under so much pressure to make up our minds, to declare our allegiance to one hypothesis or the other?

EW: What is your reaction toward the various accounts of life after death by people who had clinically died and were revived?
CS: Well, it's all anecdotal. We have people who have had near-death experiences and have been resuscitated. For all I know, these experiences may be just what they seem and a vindication of the pious faith that has taken such a pummeling from science in the past few centuries. Personally, I would be delighted if there were a life after death—especially if it permitted me to continue to learn about this world and others.

It is really quite striking. People in very different cultures, with different religious assumptions, still report remarkably similar near-death experiences about rising towards a brilliant light and having some glorious figure waiting for them. My guess is that there are just too many cases of that sort—cross-culturally homogeneous—for these experiences to be just conventional descriptions or useful figures of speech.

EW: What is your guess?
CS: My guess is that there has to be some deeper explanation. But that doesn't mean the explanation has to be what the people themselves report—that they went to heaven and saw a god or gods. In my book, *Broca's Brain* I tentatively propose an alternative explanation. It's only a speculation.

It centers on one experience that every human being shares that is truly cross-cultural: the experience of birth. You have spent nine months in utter darkness and now for the first time you have a hint of light—it must be absolutely dazzling, transforming. I cannot imagine a more spectacular transition. Usually, there is someone waiting for you, the midwife, the obstetrician, or the father.

It seems at least possible to me and to some others that in a near-death experience you reach for your earliest and perhaps your most profound experience, birth. I think the recollection of birth at a moment of near death may explain the stories of going to heaven. Incidentally, isn't the entire concept of baptism widely considered a symbolic rebirth?

EW: Where does all this leave religious beliefs that run through all cultures?
CS: The general acceptance of religious ideas, it seems to me, can only be because there is something in them that resonates with our own certain knowledge—something deep and wistful, something every person recognizes as central to our being. One such common thread, I propose, is birth. I think that the mystical core of such a religious experience is neither literally true nor perniciously wrong-minded. It may be rather a courageous, if flawed, attempt to make contact with the earliest and most profound experience of our lives.

EW: In listening to the voice of science embodied in your response to religious beliefs, I am now wondering about what you regard as the relationship between science and religion.
CS: In my view, they nearly don't communicate at all.

EW: Should they?
CS: Of course.

EW: In your view, what do they have to say to each other?
CS: I think religion has something to say to science about the social underpinnings of the enterprise of science, something about the goals of science, the human values that should always be in mind when we do science. There is also what Oppenheimer said late in his life about the development of nuclear weapons: scientists have known sin.

I also think that science has a fair amount to say to religion mainly about the nature of evidence. The idea of putting faith in an ancient argument from authority which is not to be questioned seems to me to have dire and dangerous implications for politics. I am concerned that the authoritarian aspect of religion poses real dangers for our survival.

EW: In all this, I find the deep human need for transcendence missing.
CS: You're in favor of mystification?

EW: No, I'm in favor of transcendence.
CS: What does that mean?

EW: It means going beyond the tangible, the empirical, because as long as you limit your view of the human condition to the empirical you don't satisfy deeper human needs.

CS: I don't agree. I think the things we call myths are made deeper, more relevant, and more compelling when they resonate with our natures, when they are based on truth, when they reflect external reality. We evolved in an environment where those of our ancestors who were not well adapted died off. Our scientific way of viewing the world has been selected because it works.

There is no question that humans all over the planet have a deep feeling for myths, so myths must serve some purpose—to give us some understanding of our context in a larger framework. Imagine our ancestors looking at the moon, the planets, the stars and making up stories to answer their need to understand. In many cases, the stories involved deities, such as the moon as a god.

Now is that myth about the moon deeper because it was wrong? Should we waffle, and say, "Well, if we redefine what we mean by a god, then we can still call the moon a god?" No. Let's admit that the moon is not a god and move on. It seems to me that it is a much greater achievement to understand what the moon is really about—four and a half billion years old, cratered by enormous explosions in its earliest history, a desolate world on which life never arose.

EW: So the important thing is to take the mystery out of myth.
CS: You can always go deeper. If you pick any topic and keep asking questions, you will always reach a place where knowledge runs out. Our powers are limited. They will always be. There are real mysteries enough, without inventing new ones.

EW: What do you do at that point? Become a believer, a nonbeliever or an agnostic?
CS: Why would you want to do anything but say we haven't penetrated beyond this step yet? There are painfully many cases in the history of human inquiry when people committed themselves prematurely and just got the wrong answers. In such a case, many people can suffer. We should learn from our history.

EW: But can human beings live by doubt alone?
CS: No, but I don't think that's the right way to phrase it.

EW: Then you phrase it.

CS: Let me say not what the question is, but what I think the answer is. What we need for survival is a well-tuned mix of creativity and doubt. In every subject, all sorts of ideas are proposed or should be. Some are impassioned, some are inspired, some are brilliant. But none of that guarantees they're right. Many of those ideas turn out to be just dead wrong, 100 percent wrong.

EW: Where does that leave us?

CS: You must be skeptical; you must ask for verification. If someone claims a thing happens in a certain way, you do the experiment to check it out, to see if, in fact, it works as claimed. You examine the internal coherence of the idea. You test its logical structure. You see how well it agrees with other things which are reliably known. And only then do you start accepting new ideas. This is standard practice in science. I wish it were more widely applied.

EW: Is it fair to say that you are standing on shaky ground when you stand on skepticism? After all, skepticism stands ready to doubt even skepticism itself.

CS: I don't think that's a contradiction. The mix of creativity and skepticism is at the heart of science. We can tell it works by looking at the advances science has made. We have performed practical accomplishments which would have left our ancestors openmouthed. Our abstract ideas, even our mathematical musings, have some validity; they really are connected with the external world. We made those advances by throwing away at least some trust. I'm afraid religion doesn't throw away enough trust. When there is insufficient skepticism, every idea is as good as every other. That's the same as having no ideas at all. It's essential to winnow the good ideas from the bankrupt ones, and skepticism is the tool.

EW: What then is—or can be—the link between religion and science? Are they total strangers?

CS: If you look into science you will find a sense of intricacy, depth, and exquisite beauty which, I believe, is much more powerful than the offerings of any bureaucratic religion. I would not even object to saying that the sense of awe before the grandeur of nature is itself a religious experience.

EW: What about the scientist who professes belief in God, in heaven and hell, and in formal religion?
CS: I would ask, "What's your evidence?" If he says it's a matter of faith, I would say he's forgotten the tested method of science at the moment. If he presented evidence, I would certainly pay attention.

Again, where religions teach us that we must accept, without challenge, a body of tradition, such religions are doing a very serious disservice to the human future. I think the only way to survive the next fifty years will be by seriously challenging the conventional beliefs—not just in religion, but especially in economics, social structure, and politics. If we're taught from our mother's knee that we must not challenge the conventional perceptions, we'll never get from here to there.

EW: In the final analysis, what does Carl Sagan, scientist, explainer of science, and embodiment of "creative skepticism" believe?
CS: My deeply held belief is that if a god of anything like the traditional sort exists, our curiosity and intelligence are provided by such a god. We would be unappreciative of those gifts (as well as unable to take such a course of action) if we suppressed our passion to explore the universe and ourselves. On the other hand, if such a traditional god does not exist, our curiosity and our intelligence are the essential tools for managing our survival. In either case, the enterprise of knowledge is consistent with both science and religion, and is essential for the welfare of our species.

A Pale Blue Dot
Claire Marino / 1992

From *NASA Magazine*, Fall 1992, pp. 32–33.

Q: What do you think mankind's perception of space was like before we sent spacecraft and humans up there to find out?
A: It's the most natural thing in the world to think that Earth is not only the center of the solar system, but the universe—put there for our edification or amusement. That has been the view of people all over the world, all through human history. And even today, even this minute, it is still with us. We talk about the sun rising and the sun setting. There's nothing in our language to indicate that it is, in fact, the Earth that's turning. I think that while most people had been taught that the Earth is round, in their heart of hearts they didn't believe it. And it was only after the advent of space exploration and in particular, the early photographs of Earth—mainly from manned missions—looking back and seeing it as a beautiful blue-white jewel set against the black velvet background of space, that people suddenly got a sense of where we are.

Q: Can you remember the particular shift in consciousness that occurred when mankind first saw itself from space?
A: Suddenly people were struck in the most direct way with a portrait of their planet taken from outside. Nothing like that had ever been seen in human history. Now these pictures of Earth are a kind of global icon. You see them everywhere; we have become a little bit habituated to them. But of course, the wonder is there again with each new generation.

Q: I know that for the general public it was very important for us to get that view of Earth, to have that ability to look back on it. Is there a difference in the way the scientific community feels when something like this happens?
A: Well, you know, scientists are human beings also, and so naturally they get caught up in the passions and prejudices of the moment just like everybody else. The sense of wonder in looking at Earth from space, I believe, is shared

as much by scientists as by the general public. And what a triumph for the human species to be able to step off the Earth and look back at ourselves. The most profound example, I believe, was obtained by *Voyager 2*; after passing the outermost planet, we were able to turn the cameras to photograph the Earth. A single pale blue dot. No continents, no clouds, no oceans. Just a pinpoint of reflected sunlight! All those billions of miles away. I find that a chilling, spine-tingling, exciting, perspective-raising, consciousness-raising experience. It's said that astronomy is a humbling and character-building experience. This is an example of it.

Q: I wonder if in some ways the fascination of looking back at Earth is one of the most important benefits of having gone to space.
A: It's one of the most important intangible benefits. Hard to put a price tag on it—unlike so many of the benefits of space exploration.

Q: I wonder if you could talk about humankind's urge to explore?
A: It's deeply built into us—after all, we come from hunter/gatherers. Humans spent 99 percent of their tenure on Earth, before civilization, in small nomadic groups. And so we were always exploring. Exploration is built as deeply into us as anything. Now, for the first time, we are in this circumstance of living in a world in which almost everything except the ocean bottom is explored. We've been almost everywhere, we humans. And so this urge for exploration has no outlet—except that at just this same moment, the universe has opened up to us. Now we have a much vaster arena for our exploratory propensities than the mere surface of one small planet circling one small star.

Q: What more is there to find out there? Why should we keep going out there?
A: First of all, if you buy my argument that exploring is built into us genetically, then there is no "why," any more than we ask why we humans enjoy being in each other's company. It's just the way we are. Still, I think the reasons for exploring are very concrete. The deepest one, to my mind, is that only by knowing what else is possible can we understand ourselves. If you're stuck on this planet, and you only know one style of volcanoes, or earthquakes, or climate change, or weather, or life, then you are fundamentally limited in how well you understand your own home.

Venus is a world with a horrendous carbon dioxide greenhouse effect, which has raised the surface temperature to about four hundred and seventy degrees centigrade. That's roughly nine hundred degrees Fahrenheit. That's hotter than the hottest household oven. It is important for us to understand that a massive carbon dioxide greenhouse effect can have such a consequence—we who are busily pouring carbon dioxide and other greenhouse gases into our own atmosphere. Go to Mars, and not only can't you find big forms of life, you can't even find microbes. You can't find the simplest organic molecules. Why? Because Mars has a planet-wide ozone hole. There's no ozone on Mars. The searing ultraviolet light from the sun strikes the surface unimpeded, and any organic molecules that were there get quickly fried. Isn't that information of use to us who are tearing holes and thinning our ozone layer? That's a very practical application of planetary exploration.

Sagan: Dump Environmentally Unconscious Slobs
Ponchitta Pierce / 1992

From *Earth Summit Times*, February 28, 1992.

The following is an interview with Carl Sagan, a professor of astronomy and director of the Laboratory for Planetary Studies at Cornell University. His published works include the Pulitzer Prize–winning *The Dragons of Eden*.

PP: What do you think the [1992 UN] Earth Summit conference in Rio will accomplish?
CS: The idea of heads of state meeting together on environmental issues is absolutely unprecedented. If the leaders of the major nations, not just the industrial nations but also the developing world, actually agree to specific actions to safeguard the global environment, it will be of historic importance. If they merely talk and do not commit to action, then it will be a photo opportunity.

PP: There are those who feel that the conference will be nothing more than, as you say, a photo op.
CS: The present United States resistance to the idea of carbon-dioxide emission limits—the United States being the major carbon-dioxide polluter on the planet—suggests this might very well be the outcome. The United States's view is that setting limits on chlorofluorocarbons for ozonosphere depletion gets it off the hook on its responsibility about global warming. That's the official Bush-Sununu position.

PP: With former Chief of Staff John Sununu out of the White House, will that make a difference on U.S. policy on the environment?
CS: There's no question Sununu played a major role in the United States's adopting a retrogressive policy. With him gone, maybe things will change a little bit. Of course, the fact that it's an election year might help as well. I don't see any way to predict. For example, is the United States going to be represented by its president in Brazil?

PP: How should the average citizen evaluate the conference?

CS: It potentially is very hopeful. Here is a set of global environmental crises brought about inadvertently. Who would have figured that the working fluid in refrigerators and air conditioners could pose a danger to everyone on Earth from increased solar ultraviolet radiation? Who would have predicted that the mere burning of petroleum, natural gas and coal could make profound changes in the global climate?

We did these industrial activities for perfectly understandable reasons, and we discover that we have outsmarted ourselves, that we haven't understood the fragility of the Earth's atmosphere and the power of our technology.

Cause for some pessimism is the fact that it is very hard to undo the damage. Chlorofluorocarbons stay in the atmosphere for almost a century. So, if we stop producing them altogether at this moment, there would still be significant ozone depletion in the time of our grandchildren and great-grandchildren. The battle is between short term and long term, between established industries and politics and a safeguarding of the future.

PP: Is it ironic that it seems easier to raise billions to fight a war in the Gulf than to preserve the environment?

CS: It's much easier to demonize the head of a foreign nation, especially one with a different culture from ours. It is much more difficult to raise public concern about invisible gases. And dangers that you have to understand some science to recognize, especially when the damage happens in the time of our children and grandchildren.

There are a lot of people who can't mobilize their activities that far in the future. And chief among them are politicians. Politicians are concerned about their term of office.

PP: As a human being, does that make you angry? Resigned? Frustrated?

CS: Of course, it's very disturbing to see the priorities so warped and selfish. On the other hand, most leaders themselves have children and grandchildren, as do captains of industry. There's cause for hope there.

PP: Much attention has been paid to the fact that we did underestimate the danger of ozone depletion. But there are still those who are saying that we are overestimating global warming, that the final record isn't in.

CS: Shall we wait until the evidence is absolutely compelling, namely that we are in the throes of a global environmental crisis? Or shall we take prudent steps beforehand in case the vast majority of scientists who are looking at this issue are right?

I find a strange contradiction between the attitudes of those who urge a go-slow on this issue and, in many cases, their own attitudes about what we should do about military preparedness. That we should not plan on what the adversary is likely to do, that we should plan on what in the world the adversary could possibly do. I don't hear anybody of that mental framework saying the same thing about the global environment.

PP: The dangers to the environment are so great, people might say, "What can I do on my own to help?"

CS: The major problems are not solvable by individuals. The U.S. military, for example, is a major source of carbon-dioxide emissions. So is the auto industry. It is not enough for individuals in their everyday lives to take action which can restore the balance. They must be involved in the political process.

It has to be that environmentally unconscious or unconcerned politicians are voted out of office. As soon as that happens you will see an amazing surge of environmental consciousness among politicians. It has to be that consumers boycott products—including gas-guzzling automobiles—that are environmentally dangerous. You will then discover a newfound commitment to the environment in Detroit.

PP: We have been talking about the environment on Earth. What about space?

CS: Look at the planet Venus, where the carbon-dioxide greenhouse effect has brought the surface temperature to about four hundred and seventy degrees Celsius, hot enough to melt tin or lead. I'm not saying there was once a species of Venusians who existed on driving fuel-efficient automobiles. But if anyone is skeptical about the danger of a greenhouse effect, all they have to do is look at our nearest planetary neighbor.

This is also true for ozone depletion on Mars. The Martian surface is fried by ultraviolet light because there's no ozone layer there. So the nearby planets provide important cautionary tales on what dumb things we should not do here on Earth.

Talk of the Nation: Science Friday
Ira Flatow / 1994

Transcribed from *Talk of the Nation: Science Friday*, December 16, 1994. © NPR® 1994. Any unauthorized duplication is strictly prohibited. Reprinted by permission.

Flatow: This is *Talk of the Nation: Science Friday*, I'm Ira Flatow. We'd all like to think that we're pretty much the center of attention, the center of the universe, but in the words of Carl Sagan, "We live on a routine planet near a humdrum star stuck away in an obscure corner of an unexceptional galaxy which is just one of a hundred billion galaxies in the universe." And if you think that sounds depressing, consider this: There's no guarantee that our boring little rocky planet will be around forever. If we don't destroy it, maybe a stray asteroid will. And so where does that leave us? Astronomer Carl Sagan might say it brings us back to our roots as explorers, and may drive us to become interplanetary—even intergalactic—wanderers. This hour Carl Sagan joins me to talk about the future of astronomy, space exploration, and the human race. Dr. Carl Sagan is the David Duncan Professor of Astronomy and Space Science and the director of the Laboratory of Planetary Studies at Cornell University in Ithaca, New York. He is cofounder and president of the Planetary Society, and author of the new book *Pale Blue Dot*. It is my pleasure to welcome Carl Sagan. Welcome to the program!
Sagan: Thanks very much.

Flatow: *Pale Blue Dot*. That's always the first question that every interviewer asks an author: why the title?
Sagan: Well, I was an experimenter on the *Voyager 1* and *2* spacecraft and after they swept by the Jupiter, Saturn, Uranus, and Neptune systems, it was possible to do something I had wanted to do from the beginning, and that is to turn the cameras on one of these spacecraft back to photograph the planet from which it had come. And clearly there would not be much scientific data from this, because we were so far away that the Earth was just a point, a pale

blue dot. But when we took the picture, there was something about it that seemed to me so poignant, vulnerable, tiny, and if we had photographed it from a much further distance, it would have been gone, lost against the backdrop of distant stars. I thought, "There, that's us, that's our world, that's all of us, everybody you know, everybody you love, everybody you ever heard of lived out their lives there on a mote of dust and a sunbeam," and it spoke to me about the need for us to care for one another, and also to preserve the pale blue dot, which is the only home we've ever known. And it underscored the tinyness, the comparative insignificance of our world and ourselves, as you said in your opening remarks.

Flatow: Back when men were walking on the moon, there was that famous photo of the earthrise over the moon and the—I guess you might call it the bright blue marble as compared to your pale blue dot—that sort of led to movements like the environmental movement, where people could see us as a united planet without the political boundaries.
Sagan: Exactly.

Flatow: Can we use the pale blue dot as an analogy to that, as something that's even further looking?
Sagan: That's it. It's a set of steps outward, and that *Apollo 17* picture I think raised many people to an environmental consciousness, and the pale blue dot—at least for me—represents the last moment in spacecraft leaving the Earth in which you can see the Earth at all. And the idea that we are at the center of the universe—much less the reason that there is a universe—is strongly, powerfully counterindicated by the smallness of our world.

Flatow: Whatever happened to the man in space program? I don't have to tell you how popular it was—it was the talk of the sixties; we all grew up with it. There was excitement, there was fervor, the exploration, everybody was behind it, countless amounts of money was going into it. Now it just lies fallow.
Sagan: Absolutely right. I think the first thing to say is that was a historic, mythic achievement. A thousand years from now when nobody has any idea what GATT is or who the Speaker of the House was in the late nineties of the twentieth century, people will remember Apollo because that was the time that humans first set foot on another world. But Apollo was not about science, it was not about exploration, Apollo was about the nuclear arms race,

it was about intimidating other nations, it was about "Beat the Russians!" And when we did beat the Russians, then the program was ended, and the clearest indication of that is the fact that the last astronaut to step on the moon was the first scientist. As soon as the scientists got there, the program was over. People said, "Boy, are we wasting our money on science!"

Now, lately, in the seventies and eighties and nineties, NASA has been very.... For the manned program—or human program, I hate to use the word "manned" because there are women astronatus—in the human program, we're shuttle-oriented. What shuttle typically does is put five or six or seven people in a tin can two hundred miles up in the air and they launch a communications satellite or something that could just as easily be launched by an unmanned booster, and then the newts are doing fine or the tomato plants didn't grow or now, with the next one, they're going to see how soft drinks taste in low-earth orbit, for heaven's sake. And then they come back down again and they say, "Oh, we've had another exploration." That's not exploration; that's like driving a bus over the same highway two hundred miles.

Flatow: That's cola wars in space!
Sagan: The cola wars. Whereas if NASA had gone on to send humans to near-earth asteroids, or to land on Mars, the enthusiasm would have been maintained at a very high level. Now I don't say that it's NASA's fault—NASA cannot make that decision on its own, it has to be made at a much higher level—but that decision was not made, NASA was left to its own devices, and that's why we have a falling off of interest in the space program. For excellent reasons! People aren't stupid; they understand we're not going anywhere. Now on the question that you raised at the very top of the show about isn't this terribly expensive and don't we have enormously pressing needs, of course we have other pressing needs, and that does take money. But look how the arithmetic works out: If we're not in a hurry, if we're talking about a few decades, and if the United States were to join with the other spacefaring nations on the planet, this could readily be done without any increase in the existing budgets. If we focused on the proper objectives, we could do this without breaking any banks at all.

Flatow: You point out in an interesting point that most people think that the NASA space budget is as big as our defense budget is when it's in fact only 5 percent.
Sagan: Yeah, exactly.

Flatow: People think, oh, we're spending all this money in space. When you look at the budget, we're really hardly spending anything.

Sagan: It's true. And just parenthetically, when we think of all those pressing social and environmental and other needs and wonder where to get the money from, the Department of Defense spending—including hidden costs—over $300 billion a year in a post–Cold War ear is a very good place to take a close and hard look.

Flatow: Another interesting point is that you mentioned in your book that . . . as close as July 1989, President Bush—on the twentieth anniversary of the *Apollo 11* landing on the moon—announced "a long-term direction for the U.S. space program called the Space Exploration Initiative. It proposed a sequence of goals including a space station, a return of humans to the moon, and the first landing of humans on Mars. In a letter statement, Mr. Bush set 2019 as the target date." Do you know anybody who talks about this program anymore? Whatever happened to it?

Sagan: What happened to it is it died in the process of being born, because the Republican administration was not willing to commit any political capital to get it done. It's very easy to say we'll do something by 2019—that's, whatever it is, three and a half presidencies in the future—and who knows who will be president then. You can't commit your successors. The thing about President Kennedy's Apollo program was that he made his historic speech in 1961, which said that we would use rocket boosters not yet conceived, alloys not yet invented, rendezvous and docking techniques not even conceived, to go to a moon that no one had ever been to, and we would do this by the end of the decade. And this was announced at a time no American had even achieved earth orbit, but the timescale was politically within reach—and the amazing thing is that we did it on that timescale. It is truly an extraordinary technological and human achievement.

Flatow: You make a case for colonizing space different than most people do in this book. It's an excellent book—my own personal opinion is that it's one of the best books you've put out recently. It's really very interesting to read and chock full of stuff that I think most people don't realize about space and exploration. Your tack in this book is that you argue: Let's not go out in space for things you could argue for, science, exploration, education—you argue that we have to colonize space because that's the only way we might survive in the future.

Sagan: That's right. I'm a big fan of robotic space exploration. I have been involved with it for thirty-five years. If you want to do science, that's the way to go—it's cheaper, it doesn't risk lives, you can go to more dangerous places, and so on. But as with Apollo, the only justifications that will work in the real world for human spaceflight are ones that involve some much broader political or historical agenda, and I believe there are three. One is emotional—and a lot of people feel it, I know a lot of people don't—and that is that we come from wanderers, from hunter-gatherers, 99.9 percent of our tenure on Earth was in that condition: no fixed abode. It was long before we had villages and cities, and now the Earth is all explored, we're in some sedentary hiatus, and I think a lot of people long for some exploration. You don't have to do it yourself because of virtual reality—a few people exploring can communicate it to many. On the other hand, if your child is hungry, the appeal of this argument is not very hot.

Flatow: Yet, parenthetically, when comet Shoemaker-Levy 9 smashed into Jupiter, it was front-page news.

Sagan: Of course, absolutely. And that brings me to the second and third points, which are much more immediate and practical. While I do not for a moment suggest that the Earth is a disposable planet, and I think that we have to make the most heroic efforts to preserve the environment, it is a fact that our technology has reached formidable, maybe even awesome proportions. The environment that sustains us is very vulnerable. The thickness of the atmosphere we breathe is, compared with the size of the Earth, is about the thickness of the coat of [laminate] on a schoolroom globe. And that being the case, there is a chance that we will do ourselves in. We're certainly a danger to ourselves. I would like to see self-sustaining human communities on other worlds—in the long run, there's no big hurry—so that we hedge our bets or diversify our portfolio. Clearly our chances are much greater if we do that. And the third point is there is a specific danger that we are now able to identify, and that's connected to what you just said about Shoemaker-Levy 9 smashing into Jupiter last July. The Earth lives in a bad neighborhood in space. We orbit the sun amid a swarm—an enormous number—of asteroids and comets, and you just take one look at the distribution of these orbits and it's clear that the Earth has to run into them, or they into us. Most of them are little, burn up in the atmosphere, and don't do much harm, but the longer you wait the more likely it is that a big one will hit. The ones that hit

Jupiter last July were the biggest ones there were, about a kilometer across. They produced a blotch in the clouds of Jupiter that was about Earth's size. And a kilometer across object is the size that would cause enormous environmental damage to the Earth. A ten kilometer object that hit the earth sixty-five million years ago wiped out the dinosaurs and 75 percent of the species of life on Earth. Now to deal with this, first of all, we have to inventory these near-Earth objects. Surely we should be busy finding out if there's any danger from any particular object. We're not even doing that yet. And secondly, we ought to develop a technique to deal with an errant asteroid or comet if it's found to be on Earth-impact trajectory and, without going into (we can if you want) the techniques for doing that, there's no way of doing that unless we're out there. So this is, I claim, a very practical reason why in the long term humans have to be out in the inner solar system at least.

Flatow: Scope out for us, briefly, in a nutshell, what sequence of events would you foresee for us to go out and colonize, and where would be a good spot to look to live?

Sagan: I think there's a set of steps, the first of which is better scientific exploration of other worlds so we know the lay of the land, and the development of the technology for safe survival of humans in space for long periods of time. That ought to be the principal focus of the International Space Station project that the United States is leading. It's not quite. I think it will probably be, but it isn't yet. And there's a few connected things with that—you would like to test out our ability to hide from solar flares, energetic events from the sun, you don't want to fry your astronauts. That happens not all that often. And then eventually, there's a set of objects that are accessible. Near-Earth asteroids—the very culprits we're worried about—some of them are easier to get to than even the moon, and much easier to get back from than the moon. Some of them are really strange looking, as if it's two worlds glued together, suggesting that we have here, in microcosm, or stopped motion, part of the process that led to the origin of planets. We might be able to learn about our own origins there. And because of the low gravity, we can do all sorts of engineering work there and so on. But the real test—the real focus—ought to be Mars, the nearest earthlike planet. It has an atmosphere, polar caps, winds, two moons of its own, enormous volcanoes, but most important it has clear evidence that four billion years ago it was a warm and wet world, unlike today. Four billion years ago is also the time that life arose on Earth, and is it possible that two very similar nearby worlds, life arises on one or not the other? Or did life arise on Mars four billion years ago? Might it be—despite the negative Viking results—hanging on in some refugia subsurface, some oases? Or maybe it became extinct, and the fossils—chemical and morphological—are waiting for the explorers from Earth. Mars is a very exciting place, and I would say those are the obvious objectives.

Flatow: Why don't we go to the phones to Robert in Virginia Beach.

Robert: Dr. Sagan, let me first say it's an honor to speak with you today.
Sagan: Thank you.

Robert: You're welcome. I have a question. I guess you've answered it in part, but I'll fire it anyway. Seems to me that you mentioned that we're doing a lot of things to the environment that have an unsatisfying result for us. They always say nature changes things and compensates, but we may not like what it does. So I'm wondering, with respect to colonization, where would we go? Where are the likely places we would go? What's the timetable to get there, and what are the basic steps for us to take before we can get there?
Sagan: Well, I sort of answered that just a few minutes before you called.

Flatow: Let me take it a step further and say, let's say we were to go to Mars, to take the intermediary steps and go Mars. Are we going to, as they show us in science-fiction movies, try to change the atmosphere of Mars and create giant colonies, or are we going to try to live in shelters there?
Sagan: Well, you see, the timescale I'm talking about is not in the next few years. We would start in the next few decades, and we would really get going in the next few centuries. That's the appropriate timescale for the technology. So first there would be the first human landing on Mars, an international crew very likely, carrying environments and spacesuits and returning to the spaceship overnight. That would be followed by rudimentary habitats, closed ecological systems in which you could live inside a bubble, maybe something like Biosphere 2 in Arizona. You would grow into a set of these, you can think of them as villages, but the long term—the grand possibility, and we don't know if it's possible—is to convert the surface environment of Mars into something much more benign, much more earthlike, something that science-fiction writers have called terraforming, transforming into something like the

Earth. And while this is extremely difficult to do for, let's say, Venus, it doesn't look all that impossible for Mars, at least part of the way. The key point is that Mars is too cold, and the atmosphere doesn't have an ozone layer so deadly ultraviolet light from the sun is striking the surface. Both of those mean: Put more atmosphere into Mars. And because it's cold, there's a lot of gases frozen away in the soil, chemically bound to the soil, or there as permafrost and polar caps, and there might very well be ways to release the frozen and chemically bound gases already on the surface of Mars into the atmosphere, warm the place up, and shield the surface from the ultraviolet light. We don't know that; we obviously have to do more work there. By the way, one key thing about going to Mars—if we can pull it off, it'll be much cheaper than otherwise—is to use Martian resources to generate fuel and oxidizer for the return journey. If you don't have to take your fuel and oxidizer to get back—if you only have to have enough to get to Mars, and there generate enough to get back—the weight you have to carry to Mars is much less, and the cost of the mission is *much* less.

Flatow: How would you do that? Take it out of the soil?
Sagan: One most interesting possibility, due to Martin Zubrin of Martin-Marietta in Denver, is you carry compressed methane. You combine it with the carbon dioxide from the atmosphere. You generate molecular oxygen and combine it with the methane—you generate the molecular oxygen from the CO_2—and now you have your fuel and oxidizer. And for long-term human stays on Mars, the molecular oxygen is used for breathing, the water is used for drinking and bathing, and as much as you can use the local resources, there's an enormous multiplier factor in how much you save in getting there. There are a set of clever ideas that have not at all been exploited, and it might turn out to be much less grand in terms of fiscal drain and activity than people have imagined.

Flatow: Dr. Sagan, any new TV stuff or movie stuff? I understand you're working on a film. Is that correct?
Sagan: Yes. I wrote a novel in the middle eighties called *Contact* about the first receipt of a radio message from an advanced civilization in the depths of space, and now Warner Bros. is making it into a major feature film, as they say. My wife Ann Druyan and I are co-producing and co-writing, George Miller, the Australian director, is directing, and Jodie Foster will be the lead.

Flatow: Big name!

Sagan: [laughter] Yes! Just delighted! So the summer of '96, if all goes well, it should be realized.

Flatow: Well, since you're helping and producing the film, I guess it would be true to the text, then. A lot of movies do not go along with the book version. Would we expect that to happen here?

Sagan: I would say it's a little early to be sure. [laughter] For one thing, you know, movies have a different idiom and requirement than books and especially than novels. I could spend a lot of time in a novel telling what's inside the head of a character; in a movie you've got to show it. It's a very interesting discipline, the difference between writing books and writing movies, and we've been greatly helped by Linda Obst, the executive producer, and George Miller in learning this. But so far, at least, it is true to the book, although changes, to make it filmic idiom, so that it really works in cinema, of course are being made.

Flatow: Whatever happened to the SETI project, the search for extraterrestrial intelligence? Is that moot, dead, defunct?

Sagan: Well, it's very interesting. Let me spend just a couple of minutes on that. There are a number of SETI projects. You use large radio telescopes to see if anyone is sending an intelligible message. Let me say a few words about one such study called Project META, and then I'll go on to the NASA one, which I suspect is what you're talking about. META is a program sponsored by a private membership organization—five and ten dollar contributions of members—a nonprofit organization, called the Planetary Society, that you mentioned at the top of the show that I'm president of. After five years of study and two years of followup, Paul Horowitz, who's the project director—he's a physics professor at Harvard—and I published a paper last year in the *Astrophysical Journal*. What we found is this: To discriminate genuine extraterrestrial intelligence signals from other extraterrestrial radio waves in space and from a huge radio frequency interference problem down here on earth, we used a set of discriminants, or filters: narrow-band transmission, it has to not rotate with the earth, it has to be stronger than the occasional statistical noise that all electronic systems have, and so on.

After we did that, we found there was a handful of events that passed through all the filters, and the five strongest of them, the five most intense putative signals, all came from the plain of the Milky Way galaxy. Now that's where the stars are, and you would not expect that a glitch in the electronics would only go on when you're looking at the plain of the Milky Way, and so, you know, that's enough to make the heart start palpitating a little bit and goosebumps break out, but there's something extremely odd about it, and other search programs have found the same thing: When you go back and look at these places two minutes later, it's not there. A day later, a month later, seven years later—we've done all that—we never see it again. And in science, a non-reproducible result is almost worthless. You have to be able to go back and check and have other observers who are skeptical or make different assumptions than you check it out. So we don't know what that is, certainly those places in the sky deserve further examination, and we are moving on to a much bigger project called BETA, billion-channel extraterrestrial assay, which Paul Horowitz has almost ready.

Now, at the same time, a still more sophisticated program was supported by NASA. It went on the air in October 1992, funded by Congress, and was ignominiously turned off by Congress just one year later. The argument presented by Senator [Robert] Bryan of Nevada was that we didn't really know that there could be extraterrestrial life out there, and also it was too expensive. Well, of course we don't know whether there could be—the whole point is to find out! If we knew beforehand, we wouldn't have to look. And the consequences of success are enormous—I mean, the transforming—it's hard to think of a more important discovery. And as far as cost goes, the NASA-SETI program was costing about one attack helicopter a year. Now there's a very nice coda to this story, and that is that while NASA is not supporting it, a number of captains of the electronics industry have made contributions totalling something around seven million dollars so that the project is going to go back on the air in Australia sometime early next year, and that's something really great. The search program is so important, and the technology is now sufficiently inexpensive, that this could go on even without government support—but it sure would be great if the government would change its mind on this.

Flatow: John in Elsa, Illinois. Hi, John!

John: Hi, Ira! Dr. Sagan! It's good to have the chance to talk to you!
Sagan: Thank you!

John: Dr. Sagan, this is a little bit off the subject, but earlier in the show you were talking about the environment that sustains us on the planet, and our technologies advancing to the point where we can become a danger to ourselves. There's a discussion that I've heard raised a few times within the past couple of years about a solar energy collection system that would either be in an orbit of some sort—maybe geosynchronous—or mounted on the moon, and would beam, I think by microwave energy, power back to the planet.
Sagan: That's right. Yeah.

John: Seems to me that this could possibly be creating an effective increase in the size of the disc that Earth presents to the sun, and as such couldn't that be raising the net energy collection by the planet, creating some sort of a second cousin to global warming?
Sagan: I don't quite see how that would work, but, you know, I don't use that argument in my book *Pale Blue Dot*. Why not? Because if you have a means of converting sunlight into electricity, why put it in Earth orbit? The arguments are "okay, well, you can put it high up enough that you're always looking at the sun." That's just a factor of two. The expense of putting it up into Earth orbit—and then beaming the energy down with microwaves—is much more than a factor of two. This scheme—which was looked at by the Congressional Research Office and by the National Academy of Sciences—does not seem to be cost-effective. However, the general question you raise—if global warming is produced mainly by greenhouse gases, the burning of fossil fuels, coal, oil, natural gas, wood—can we find some alternative energy sources, the answer is absolutely. We can do the conversion of sunlight into electricity on the ground, we can use wind turbines, we can use biomass conversion, we can use hydrogen fuel cells and, with any serious development of that technology, we can gradually displace the fossil fuel economy and, before then, we can use the fossil fuel economy much more efficiently. Why are we content with cars that go twenty-five miles a gallon when it's perfectly possible to have cars that go seventy-five miles a gallon, with adequate acceleration and looking spiffy and safe? It's perfectly possible to do. There are many things we can do with down here technology to make our environment much safer.

Flatow: Okay, let's go to Shawn in Kansas City, Missouri. Hi, Shawn!

Shawn: Hi! What I wanted to ask Dr. Sagan was it seems to me that the way to increase space exploration is to show commercial industry that it would be profitable to do so, because I think as soon as you show business that there's money to be made in space, you'll have to fight to keep them on the ground, and I just wanted his comments on that.
Sagan: Thanks! I think you're absolutely right—if there was money to be made, you'd have to fight to keep them off.

Flatow: You mentioned one of the reasons you might go into space is there might be diamonds—
[laughter]
Sagan: Wait. Before we get to that, which is essentially a science-fiction theme—why is it that industry is not elbowing each other to get into space? And the reason is that there is no commercially viable project that anyone has come up with, except of course for the aerospace manufacturers who have something to do by building the means to get up there. But no crystals, no pharmaceuticals, no ball bearings, no alloys of admissible metals, nothing like that. The criterion ought to be this: To make your technology in space is going to cost X dollars. Can you produce a cheaper or better alternative product down on Earth for X dollars? And the answer always seems to be yes. When the answer is no, we'll have industrialization, but it's possible the answer will never be that it's cheaper to do it up there. Now, there are some exotic possibilities, and Ira just mentioned one, and that is that there is a single paper in the Japanese scientific literature suggesting that diamonds might be naturally made on Mars more naturally than on Earth, okay, so maybe.

Flatow: Now we've got your attention!
[laughter]
Sagan: And in that case we can have General Electric and De Beers finance the space program, but you can't be sure of that, and in any case we have to go to Mars to find out.

Flatow: Yeah, but to come full circle on this, you're making the argument that the reason we have to go into space is not for commercial reasons but for solely practical survival reasons, that the odds are better—and you said it, but I can't remember where—the odds are better of an asteroid or comet smashing into our planet and destroying us, or dying in such a collision, than dying in an airplane crash.

Sagan: It's like this. As far as we can tell from the present statistics of near-Earth asteroids, we can ask what is the chance that the Earth will be hit in the next century by an asteroid or comet that will destroy the global civilization. I mean, that's the right question. And the present answer is one chance in two thousand. Now, you can decide whether that's a large number or a small number, but by comparison the chance of dying in a single, randomly-selected, commercial, scheduled airline flight is one in two million. And now, a lot of people worry—especially these days—about flying in airplanes, and they take out insurance policies. All I'm saying is here, also, with the odds a thousand times higher, we should take out insurance policies.

Flatow: Michael, hi, age ten! Michael, how are you?

Michael: I'm fine. My question is now, if you could get to the center of the galaxy, the Milky Way, what would it look like, could you colonize it, and how would get there, by what type of ship, what type of engine?

Sagan: Really good questions, Michael, and I'm so glad at age ten you're that far along! By age twenty, I hope you will be making significant contributions to the subject. The center of the galaxy is about twenty-five thousand light years away. If we could travel almost at the speed of light—we can't travel at the speed of light, but if we could travel almost at the speed of light—then on board the ship, it could take us very brief periods of time to get there, but as measured from the earth, it would be twenty-five thousand years for us to get there. So if you went there and fiddled around a little bit and came back, it would be fifty thousand years later and all your friends would be gone.

So that is a requirement imposed on us by special relativity—it's a law of nature, and it looks very hard to get around that except for an enormously advanced civilization with much more powers than we have, and I talk about that in that novel, *Contact*, we were talking about for. What it would look like—well, you see, we live out in the boondocks of the galaxy, and it's dark because the stars are so far apart. At the center of the galaxy, the stars are much closer together, and it is gorgeous—multicolored stars, I wouldn't say touching, but very much closer together than they are here. The idea of making human communities at the center of the galaxy may be, but that's a dangerous place, the center of the galaxy—it blows up every now and then and it looks as if there is a giant black hole at the center of the galaxy. I think we ought to stay for a while out here in our remote spiral arm where things are a lot safer.

Flatow: You want to be an astronaut or an astronomer when you grow up?
Michael: A scientific engineer.

Flatow: Okay! Good luck to you!
Michael: Bye!

Flatow: Thanks for calling. Bye!
Sagan: That was great! That's wonderful!

Flatow: We get a lot of young callers on *Science Friday*, and we're very happy to invite them to call. I guess sometimes they're home early on Friday from school or wherever—I don't care if they're playing hooky listening to our program, that's just fine. [laughter] You know, one of the most interesting parts of the book—and you have it right at the beginning, toward the front—is most of us, when we think about where would we like to find the origins of life in our solar system that would be similar to the way it evolved on Earth we'd say, "Let's go to one of the planets, go to Mars, go to Venus." But you—because you have studied this for a long time—say, "Let's go to a moon of Saturn called Titan. That's where we may find those primordial building blocks of life." Why Titan? What's going on there?
Sagan: Yeah. It's such a great finding, and so unexpected, who would have figured—just as you said, you would have figured Mars or nearby. Titan is the big moon of Saturn, and it's covered with an orange haze layer and clouds. That's really weird for a moon to have clouds and an atmosphere. Not just that—the atmospheric pressure is the closest of any world in the solar system to what it is here, and the atmosphere is made mainly of nitrogen, N_2, just as the atmosphere of the Earth is. Now, what is that orange stuff? We know now quite reliably—and I think we can really be almost confident about it—that it is complex organic matter including, if you drop it in water, the amino acids, the building blocks of proteins, and the nucleitide bases, the building blocks of the nucleic acids—the very stuff of life here on Earth, and it's dropping from the skies like manna from heaven—

Flatow: But it's cold.
Sagan: Absolutely right. So some of the building blocks—key building blocks—are being made, and are being preserved, you would think, because of the very low temperatures, so they don't decay, they're waiting for us, let's go find them. But it's even better than that. The Saturn system is of course much further from the Sun—ten times further from the Sun than the Earth is—so it has to be very cold. It's 94 Kelvin or something like that on average at the surface of Titan. And so you would say, "Look, this is the place where it misses out being an analogy with the Earth because you have liquid water here, that's essential for life, they don't have it there." But we know that the solid surface of Titan contains ice, and when a comet slams into Titan, it produces a temporary pool and slurry of liquid water. So now we can ask over the whole history of Titan—for an average place on the surface—how long did it see liquid water, and the answer seems to be something like a thousand years. A thousand years in which the organics that fell from the sky are mixed in with liquid water at reasonable temperatures. Is that enough to make a significant further step toward the origin of life? We don't know. But Titan is sitting there, waiting for us, and we're going, because in three years a joint NASA/ESA mission called Cassini is to be launched to arrive in the Saturn system in the year 2004 and an entry probe, capable of examining organic chemistry, is going to enter into the atmosphere of Titan, sampling as it descends and, if we're lucky, it will survive the landing and see what's down there. It's a very interesting fact that if you want to understand about the origin of life on Earth, the best place to go may be Titan.

Flatow: Amazing. And of course a lot of this came out of the Voyager, mostly all of it, the modern stuff we know came out of the Voyager missions.
Sagan: Quite right. And the Titan stuff I've just been describing is fundamentally based on Voyager data. You see, there's a spacecraft—two spacecraft, *Voyager 1* and *2*—product of American industry, run by the government via the Jet Propulsion Laboratory of NASA and Cal Tech, that came in on time, under budget, and vastly exceeded the expectations of its designers. It is responsible for almost all we know about most of the solar system—the Jupiter, Saturn, Uranus, and Neptune systems—and now, those two spacecraft, still working splendidly, are on their way to the stars.

Flatow: Looking back at that pale blue dot. Let's go to Jane in Eugene, Oregon. Hi, Jane!

Jane: I'm wondering . . . Mr. Sagan, if you were assuming that we would not trash any new environment that we may create out in space, and if so what do you base this assumption on?
Sagan: This wildly optimistic assumption.

Jane: Yes!
[laughter]
Sagan: Of course we are a lot more slovenly than we ought to be, and we are not doing well with our own planet, and you might very well argue: Let's hold off messing up other worlds until we can demonstrate we know what to do with our own. Let's make the Earth an Earthlike planet before we talk about making other worlds an Earthlike planet. I would be very concerned along these lines if there were life on some other planet. Then I would say that planet belongs—whatever the word belong means—to the beings on that planet, and we have a real responsibility to exercise the most extreme care there. But as far as we know, there is no life in the entire solar system except on the third planet from the Sun, the Earth.

Flatow: Is our society ready for news about life on another planet, if we were to conclusively say we have discovered life someplace else? Can we handle that?
Sagan: If it's microbial, I think nobody is going to worry about it at all. But if we get a message from another civilization in the depths of space, that's very different—and I try to imagine what the various reactions of various human constituencies will be in my novel *Contact*. I think many people would look at it with an enormous sense of wonder. You see, if we got a message, it would have to be from somebody much smarter than us, because anybody dumber than us is too dumb to send a message—we've just invented radio. So really smart guys telling us what they know. That means that every branch of human knowledge is now up for reconsideration. Some people, of course—and not just human knowledge, but things like social organization and religion—some people of course will be defensive about it, and will worry, what have they assumed that isn't true, and even in science, you know: Did we get something wrong in fundamental astronomy? Did we make a mistake in mathematics somewhere? You can see people being really nervous, but the chance to tap into such knowledge—it's like going to school for the first time.

Flatow: I'm running out of time, I have just two minutes left, but while I have you here I have to ask you a couple of science questions. One, what is your take on the problem that we've just been listening about, the news that the universe may be younger than some of the galaxies?

Sagan: It's fantastic, isn't it? It's like someone telling you that their children are older than they are—you know something is wrong. But we're just talking about factors of two: Either our method of dating the stars is wrong, or our method of dating the universe is wrong. Those are the only two possibilities. I think the most likely case is that we have the age of the stars right, and will find out that there is something wrong with our dating of the universe. But tune in! It's a great question.

Flatow: And the other great question is: What is all the missing dark matter? Do we have any idea? It gets worse all the time, the more we keep hearing more about it.

Sagan: Well, yeah, there are plenty of ideas, and all mutually exclusive. Dark matter is just stuff that we know is there because of its gravitational influence but we can't see. Well you, Ira, and I are sources of matter that don't radiate much into space—and yet, we have some mass—it might be, you know, snowballs. It might be neutrinos with rest mass. It might be black holes. It might be a kind of elementary particle that no one on Earth has detected yet. We don't know. It ranges from the prosaic to the extremely exotic and there, too, we're going to find out the answer.

Flatow: It's very sobering that we could be sitting in objects that 95 percent of the universe is made of and we have no idea what it is. That is really a sobering thought.

Sagan: In that way it's depressing. But the other way is, look, we've discovered that they're there, and now let's find out what it is, and we are on an upward trajectory towards learning, and hats off to science for figuring that out.

Bringing Science Down to Earth
Anne Kalosh / 1994

From *Hemispheres*, October 1994. Reprinted courtesy of *Hemispheres*, the magazine of United Airlines.

As a boy growing up in Brooklyn, New York, Carl Sagan gazed in wonder at the night sky. A trip to the library for a book on stars sealed his fate: He would devote his life to astronomy. Sagan recalls his college years in the 1950s as a time of tremendous optimism about science and the future. After earning a Ph.D. in astronomy and astrophysics from the University of Chicago at age twenty-five, he taught at a constellation of prestigious institutions before joining Cornell as David Duncan Professor of Astronomy and Space Science. He also directs the university's Laboratory for Planetary Studies.

Sagan has played a leading role in the Mariner, Viking, Voyager, and Galileo spacecraft expeditions and is noted for research in such areas as the origin of life, the greenhouse effect on Venus, and the long-term consequences of nuclear war on Earth. This year he received the Public Welfare Medal, the highest award of the National Academy of Sciences. He is cofounder and president of the Planetary Society, the world's largest space interest group, and a founding fellow of the Committee for the Scientific Investigation of Claims of the Paranormal, which analyzes—and debunks—psychics, channelers, astrologers, and reports of UFOs and alien abductions.

"Billions and billions of stars" has been Sagan's signature phrase since his Emmy and Peabody award-winning *Cosmos* TV series aired in sixty countries and was the most widely watched series in the history of U.S. public television. The accompanying *Cosmos* book became the best-selling science book ever published in the English language.

A Pulitzer Prize winner, Sagan boldly takes controversial stands. He was an early advocate of multinational cooperation in space exploration, a leader in the scientific community's protest of the Reagan administration's Strategic Defense Initiative (Star Wars), and was twice arrested at demonstrations against continued U.S. nuclear testing when the Soviet Union was abiding by a testing moratorium.

In his crusade to popularize science, Sagan has edited or authored twenty-five books, including *Comet* and *Shadows of Forgotten Ancestors: A Search for Who We Are*, with his writer wife Ann Druyan. His *Pale Blue Dot: A Vision of the Human Future in Space* will hit bookstores late this year [1994], to be followed by a study of skepticism and the nature of science. Druyan and Sagan are teaming up on a novel, a love story, before penning the sequel to *Shadows*.

At fifty-nine, Sagan is deeply concerned about the future of a world where the Scientific Literacy Index reveals that 94 percent of U.S. citizens are scientifically illiterate.

Q: How does not understanding science cripple people in their daily lives?
A: We live in a society absolutely dependent on science and technology and yet have cleverly arranged things so that almost no one understands science and technology. That's a clear prescription for disaster. Every day there are decisions being made in Washington that will affect our future, things like information superhighways and reduction of nuclear arsenals, research on AIDS, whether drugs that ease the pain of those who are dying should be decriminalized, what is the best way to make sure that America continues as a leader in industrial technology, how to deal with the depleting ozone layer, and global warming. There's hardly an aspect of modern society that doesn't depend on intelligent decision-making in science and technology. We are supposed to be a democracy. The people are supposed to make sure that their representatives vote correctly. How can they do that if they don't know what the issues are and can't understand them?

Q: Why aren't people keeping up with science?
A: There are a lot of reasons. In the short term we can talk about why is the basketball coach teaching chemistry, why do school bond issues repeatedly get voted down, why are teachers relying on textbooks and not labs, why is the teacher just one lesson ahead of the kids, why does the teacher discourage searching questions, when is the last time we heard some discussion of science on the Sunday morning white male pundit shows, when's the last time you heard an intelligible scientific remark by a president of the United States, when was the last television fiction series in which the hero was someone who was devoted to finding out how the world works? But these are all symptoms,

not causes. The causes, I think, are in the following: science is hard, science does not always conform to our wishes, science does not always reassure us, science puts enormous power in the hands of some people who we have every reason to mistrust. Scientists are responsible, in a certain sense, through engineering, for depletion of the ozone layer, global warming, Agent Orange, and all the rest. Now many scientists would say, "Wait a minute. We're just doing our job. All of this is the misapplication of science by government and industry." To some extent, it is. And to some extent scientists have been very courageous in calling attention to the dangers of these technologies. But, nevertheless, if we didn't have science, we wouldn't have these problems. But we also would have life spans of twenty-five years, infant mortality would be enormous, and many things that make life pleasant or even possible would be gone. There is a kind of tradeoff. All that has happened so fast that many of us are huffing and puffing to keep up.

Q: Nancy Reagan consulted an astrologer, channelers have enormous followings, and the tabloids are rife with reports that sex-crazed space aliens are abducting humans. Does it seem that there's an explosion of ignorance today?
A: No, I think we've always been like that. We had demons from ancient Greece, gods who came down and mated with humans, incubi and succubi in the Middle Ages who sexually abused people while they were sleeping. We had fairies. And now we have aliens. To me, it all seems very similar.

Q: But the difference is now we have much more knowledge and much better communication systems.
A: Well, do we? Look what's on television. How much critical science is there and how much credulous superstition? I think you could argue that television works just the other way, to make people more credulous and less critical.

Q: What can people who are afraid of science do for their children?
A: The most important thing is not be frightened when their children ask them a question to which they do not know the answer. It's all right to confess that you don't know the answer to a question, even if it's a six-year-old who's asking. The worst thing is to ridicule the child. That convinces the child there's a set of questions that adults get mad at and, after a few experiences, the child doesn't ask the question anymore; we've lost another person who might be comfortable with science. It's self-propagating. Those who are

ignorant and fearful of science create a new generation of people ignorant and fearful of science. So if you don't know the answer, you can say, "Let's look it up. Let's go to the encyclopedia." If you don't have an encyclopedia, go to the library. If you don't want to do that, you can at least say, "Maybe nobody knows the answer to that question. Maybe when you grow up, you'll be the first person to know the answer." That's an encouragement.

Q: What's wrong with teaching both creationism and evolution in the classroom?
A: Creationism is not science—it's mysticism, it's religion. I would have no problem with teaching creationism in mythology courses, in courses on social trends, even in courses on religion, which I think might be a very good thing to have. But not courses in science, because it's not science.

Q: You've been a professor for more than thirty years. How have students changed in that time?
A: The best students haven't changed much at all. They are still terrific. Beyond that, anything I would say would be anecdotal. I thought I saw in the Reagan years kids much less disposed to ask searching questions, especially of those in power. Today I seem to see a greater willingness to ask searching questions, all to the good. In the Reagan years, I thought I saw a lot more kids who chose their careers in order to make money and comparatively few who were idealistically motivated. Today, I see some signs of that turning around. But again, I could be wrong. It's not a statistically significant survey.

Q: It sounds like you think the Clinton administration is providing a better climate for science.
A: Yes, maybe a little. But certainly not enough on the environment; it's nothing like what one might have expected from Al Gore's book. The administration says, "Look, we've only been in office a year, give us some time." I'm willing to do that. Certainly in terms of knowledge of scientific and environmental issues there hasn't been a president or vice president as knowledgeable as Al Gore in decades, maybe centuries.

Q: How would you grade the administration on environmental matters?
A: I think they're very cautious, afraid to offend business. But business is precisely part of the problem because if it affects short-term profits, business

isn't concerned about environmental consequences, by and large. There are some exceptions. But more typical is the response of the DuPont company when scientists discovered that CFCs [chlorinated fluorocarbons] are dangerous: It took out ads saying, "No, no, this is only a theory. Don't worry about it." So the idea of cleaning up the environment without putting pressure on industry is naive, I think. Industry can be prodded with carrots and with sticks.

Q: Rather than cracking down on business, isn't it important for individuals to make a sacrifice? Wouldn't a gas tax that discourages people from driving be more effective in tackling pollution?
A: The way things work is, all used cars pollute a lot. The ones that might be efficient in reducing pollution, that might get more miles to the gallon, are new cars. Poor people cannot afford new cars. So as soon as you say that there's a penalty for driving cars that pollute, the penalty works preferentially against poor people.

Q: If you were the president, how would you allocate the budget?
A: Very hard question. Just think how many lines there are in the federal budget. One thing I would say, though, is to have a so-called defense budget that, including hidden costs, is over $300 billion a year when there are so many other pressing national needs is a serious mistake. The Soviet Union has collapsed. The Cold War is over. Presumably we're not obliged to invade lots of other nations. We can protect ourselves for a fraction of that $300 billion, and the money saved could do an enormous amount to solve many of our other problems. But this administration is not inclined to go in that direction.

Q: Many people fervently believe they have seen UFOs, and some claim they have been kidnapped and sexually assaulted by aliens. Do you think alien spacecraft have visited Earth?
A: Having extraterrestrials visit this planet would be great . . . even if they were short, sullen, grumpy, and sexually obsessed. Still, if they are the harbingers of advanced civilization and they're here, for heaven's sake, let's find out about them. But the thing is, the evidence is poor. In none of these cases has anyone torn out a page of the captain's log or scraped off a piece of exotic alloy of isotopic composition not known on Earth. In the abduction

paradigm there's this very interesting circumstance in which people say the extraterrestrials have implanted a little monitoring device up their nostrils. Great! Let's get one of those, and we can solve this. Now the story that comes from the abduction enthusiasts is that a lot of times people will have their implant fall out—clunk—*and they throw it away.* Think of how incurious these abductees are, not to realize this is something that can prove their case.

Also, the women who are said to have been impregnated by alien sperm. Can we take a look at the amniocentesis? How about the sonograms? How about the cases where they are born or miscarried? What are we to imagine, that the obstetrical interns would look at this half-human, half-alien baby and then incuriously go on to the next case?

Q: Is there a single piece of scientifically defensible evidence that leads you to conclude there is life on other planets?
A: Not really. You can make a plausibility argument, something like this: there's a whole lot of stars and a whole lot of planets. The biological building blocks of life are everywhere. And there's nothing in the evolutionary process unique to the Earth; Darwinian natural selection should apply anywhere. Put all that together and there is, I think, a pretty strong plausibility argument for extraterrestrial life. But that's all it is—a plausibility argument. It says it is not so absurd that we shouldn't look; it's worth looking for. So far, with robotic spacecraft—Viking, for example, on Mars—and with the use of large radio telescopes to see if anyone is sending us a message, despite some enigmatic findings, there is no compelling evidence for extraterrestrial life. But we're at the early stages of looking. We haven't found it yet. Maybe we'll never find it; maybe we'll find it tomorrow. What we need here is a tolerance for ambiguity. It's not required that we make up our minds this minute. It's perfectly all right not to pretend we know one way or the other before the evidence is in.

Q: What is the significance of the Hubble Space Telescope?
A: Here's the first big optical telescope above the Earth's atmosphere and everywhere it looks it makes major discoveries. When the telescope is turned toward the Orion Nebula, which we know is a spawning ground for stars, Hubble finds that half the stars examined have flat discs of gas and dust surrounding them. This is exactly what the people trying to understand the origin of our solar system predicted: the so-called solar nebula. It was first proposed by Immanuel Kant and by Pierre Simon, back in the eighteenth

and nineteenth centuries, just from physics. And now we see these things. It now looks as if planets are a frequent if not invariable accompaniment to star formation. There are four hundred billion stars in the Milky Way galaxy, and if many or most of them have planetary systems, the chances of extraterrestrial life are greatly enhanced. That hardly proves that extraterrestrial life exists. It's just more support for the plausibility argument. But it's tremendously exciting. And if we start looking at planetary systems, we're bound to understand more about how our own was formed.

Q: After Hubble, what's the next logical step in space exploration?
A: Hubble is an observatory in Earth orbit looking out, and that's only one aspect of space exploration. But in that category, the next obvious mission is called AXAF—the Advanced X-Ray Astronomical Facility. That's a large telescope that does just what Hubble does but within the X-ray part of the spectrum. There are energetic objects all through the universe that are invisible in ordinary light, such as black holes, that we can best understand with something like AXAF. But that's only one part of the space program. Looking at Earth to monitor its environmental health, exploring asteroids, comets, planets, moons, the sun directly with robot probes—those are some of the other things that are in the future of space exploration.

Q: What's the most serious problem facing the Earth?
A: There are lots of them: ignorance, ethnocentrism and xenophobia, population growth—although it is starting to shallow, but not nearly fast enough. And the absence of an understanding of the virtues of democracy. I would certainly include those on my list. It might be a much longer list, but those are some of the things near the top.

Q: I would have expected you to say global warming.
A: I would put that in the ignorance column.

The Charlie Rose Show
Charlie Rose / 1995

From *The Charlie Rose Show*, January 5, 1995, transcript #1285. Reprinted by permission.

Charlie Rose: Carl Sagan is here. You know him as a distinguished astronomer. His program on public television, *Cosmos*, was one of the most watched public television programs ever, perhaps—I don't have the numbers here. Perhaps *Civil War* had more viewers in attendance, but *Cosmos* was a terrific series and got a lot of attention for Carl Sagan. He has a new book, which is called *Pale Blue Dot*. It is a vision of the human future in space, and that is our subject this evening. I want to turn to this picture to get some sense of what a pale blue dot means. This is taken from, Carl—welcome to the broadcast.
Carl Sagan: How are you, Charlie?

Charlie Rose: This is taken from *Voyager 2*, is it?
Carl Sagan: *Voyager 1*, actually.

Charlie Rose: *Voyager 1*. But take a look at this, and you can see here, what? What's the meaning of this?
Carl Sagan: Well, here's this spacecraft that has flown by the Jupiter, Saturn, Uranus and Neptune systems and is on its way, astonishingly, to the stars, a triumph of human engineering. We turn the cameras back and take a photograph of the planet from where it came. And we can barely see it. Here it is, a fragile, delicate, pale, blue dot, and that's where we live. That's where every human has ever lived, and you can see the vulnerability at a glance. And that gives a humbling, and I think character-building, sense of where we are.

Charlie Rose: Humbling because it says that we are only one small portion of something that is enormously large?
Carl Sagan: Absolutely. And let me say a word about what that is. This dot is one of nine planets that goes around a humdrum star, that lives at the outskirts

of a perfectly ordinary galaxy, which is composed of 400 billion other stars, in a universe of maybe 100 billion galaxies, and recent thoughts suggest this universe is one of a very large number, perhaps an infinite number of other closed-off universes. In that context, what is the chance that we are the center or the point of the universe?

Charlie Rose: To you, no chance.
Carl Sagan: To me, no chance at all.

Charlie Rose: You also believe—I'm getting ahead of myself here—you believe that we may very well—because of what we will be able to do—that we may be able to in a sense inhabit other places because of technologies that will be able to inject onto those places certain qualities that are necessary for the survival of life.
Carl Sagan: Of human life. The technology is a double-edged sword, and its powers are absolutely extraordinary, and the rate of increase of those powers is dazzling. And not tomorrow, not in the next few decades, but on the time scale of centuries, the possibility of altering the environment of an entire world so humans could live there fairly comfortably without heroic life support equipment, seems to be really possible. And long before that, we can visit and we can establish base camps.

Charlie Rose: Yeah. You have always been—if I'm correct, and I think I am—a strong proponent of manned exploration.
Carl Sagan: Not really—I've had a very conflicted sense of it. Maybe I can say why. It's been advertised as essential for science, but it's not. Robots can do it at 10 percent of the cost, and you don't risk human lives, and the standard set of justifications for human space flight I have found wanting. But in *Pale Blue Dot*, I have come to a different opinion, that all of my objections are short-term objections, that in the long term, it is important for us humans to be out there. And I can quickly say what the reasons are. The first is we are an exploratory species. The last ten thousand years we've been sitting around in civilization. Before that, for the last few hundred thousand years, we were wanderers, explorers, nomads. And that is in our blood. And space flight is an opportunity—the only one open to us—to continue that long human tradition. Secondly, that technology I was talking about can pose a danger to ourselves. We inhabit a very thin protective atmosphere. Our technology can

destroy that environment that protects us. I don't for a moment say that the Earth is a disposable planet. We have to make the most heroic efforts to preserve it and us. But I still think it would be a good insurance policy, hedging our bets—or as Republicans like to say, diversifying our portfolios—for there to be humans on other worlds as well as here. And finally, there is a specific, identifiable hazard that—again, not in ten years, but in hundreds or thousands—starts to become worrisome, and that is that the Earth will be hit by a large asteroid or a comet, and if we're ever going to deal with that, we have to be in space. Put all that together, and I see on a time scale of decades to centuries, we really have to have a significant presence in space.

Charlie Rose: You weren't in favor of the space shuttle?
Carl Sagan: The space shuttle puts five, six, seven people into a tin can, shoots it up two hundred miles, and they launch a communications satellite that could have been launched by an unmanned booster, and then the newts reproduce nicely or the tomato plants don't grow, and then they come down again, and NASA calls it exploration. That's not exploring.

Charlie Rose: Yeah, but is it possible—and I'm way the hell out of my league here—but is it possible that they learned things from that that would enable them to do the other things that are much more important?
Carl Sagan: You're not out of your league at all. It's an excellent question. If we were into long duration space flights, space flight of about a year or something like that, as the Russian program has been, then yes, you could say, "We're learning how to go to the planets."

Charlie Rose: "These are the building blocks to go somewhere else."
Carl Sagan: But if we just go up for a week, we learn nothing about that. And this is one of many reasons, by the way, why a joint, multinational exploratory program involving Russians and Americans and Europeans and Japanese would make a whole lot of sense. Every one of these nations has capabilities that the other doesn't.

Charlie Rose: How about a space station?
Carl Sagan: Again, what's it for? The standard explanations—make money, make products that you can't manufacture down here on Earth competitively, do science, medicines—

Charlie Rose: How about medicines and other kinds of benefits?
Carl Sagan: There is not a single one of those justifications that stands up to close scrutiny. The critical question is if you were to spend the same amount of money that you were proposing to spend up there down here, could you produce a competitive or superior product? And the answer always is yes. But if our objective is to prepare for long-term human exploration in space, then space stations could start making sense.

Charlie Rose: Did we learn anything from Apollo other than the fact that we can get there and perhaps that told us something about exploration of other planets?
Carl Sagan: Apollo was about the nuclear arms race and beating the Russians and intimidating other nations. That's what Apollo was about.

Charlie Rose: National pride.
Carl Sagan: National pride, if you wish, but mainly using rocket prowess, demonstrating we had it. But as a subsidiary, as an accidental by product and advantage, there's that whole gorgeous series of exploratory missions: the Mariners, the Voyagers, the Vikings, Galileo. And on the Russian side, likewise, which have just flocked through the solar system.

Charlie Rose: And so what have we learned from those? I mean, I'm now switching to the other side. The first part of the book, you talk about some sense of our place in the universe. And then you also talk about what have we learned in the last thirty years all from the Vikings and the Voyagers and all of that. What have we learned?
Carl Sagan: This is the first moment in the entire history of the human species when we've explored first-hand—I mean not humans, but our machines, our robots, sending back the data—the environment we live in. We have examined every planet from Mercury out to Neptune. We have examined seventy moons and some comets and some asteroids. Never before has that been done, and there is only one first time. That's our generation.

Charlie Rose: But what has it taught us?
Carl Sagan: It's taught us—you take a look at Venus, you see a world with a surface temperature of 900 degrees Fahrenheit, produced by a massive carbon dioxide greenhouse effect, and never again will you be tempted to

believe radio talk show hosts who say that the greenhouse effect is something invented by the liberals.

Charlie Rose: Somebody's imagination. It is not a danger.
Carl Sagan: You look at Mars, and you see a planet without an ozone layer, in which ultraviolet light from the sun has, in effect, fried the surface so that even organic molecules cannot survive there, and never again will you say that there's no danger to depleting our ozone. We learn about our world by examining other worlds.

Charlie Rose: Is there any commitment, in terms of a national will, reflected among the public, which has to support this kind of thing? Is there any enthusiasm for exploration?
Carl Sagan: The key question is exploration. When the polls are put in terms of real exploration—not driving a truck two hundred miles up but going to some new places—the support is overwhelmingly positive, much stronger than it was.

Charlie Rose: But I don't hear politicians talking about it.
Carl Sagan: They don't.

Charlie Rose: I mean, here we are about to launch, over the next couple of years, a great debate about the role of government. Clearly, exploration is something that has to be done, I assume by government.
Carl Sagan: It has to. It's too expensive to do by private industry or wealthy individuals. What we're talking about, the advantages that accrue, are largely long-term advantages.

Charlie Rose: Long term meaning over the next hundred years, two hundred years.
Carl Sagan: No. Even just decades. And here, as in many other areas of our society, we have this fatal conflict between the short term and the long term, and it's always so tempting to say, "Let the long term take care of itself. I get re-elected on the basis of what I do in the short term." And that is extremely dangerous. It's very important, of course, to plan things in the short term, but we have to have a mix. Every great society does that.

Charlie Rose: What's been the single most exhilarating moment in the exploration process for you?
Carl Sagan: Oh, goodness. There have been so many.

Charlie Rose: Has there been one or two moments in which you said, "My heart pounded faster than it ever had?"
Carl Sagan: I must say, every time we go to a new world my heart pounds, but Viking, when for the first time we set down on Mars, where no one had ever been before, and took pictures of this landscape that didn't look the least bit exotic. It looked like Arizona or Utah. That said to me something about the commonality of processes, about other worlds having something—some similarity to our own. Another one is Titan, the big moon of Saturn, where the stuff of life, organic matter, is raining down from the skies like manna from heaven. Want to know about where we came from, where life on Earth came from? Go to Titan, the early steps are happening right now. And then, the idea of the Voyager spacecraft achieving escape velocity from the sun on their way to wander forever among the stars.

Charlie Rose: I want to turn to a couple of things. You were a very strong part of the movement against nuclear weapons, number one.
Carl Sagan: Yes.

Charlie Rose: You were very strong in your opposition to Star Wars.
Carl Sagan: Yes.

Charlie Rose: Now you begin to hear in the political community, some notion, "Well, maybe Star Wars is possible." I mean, not Star Wars, but maybe—
Carl Sagan: Missile defense.

Charlie Rose: —SDI is possible.
Carl Sagan: Well, I mean, basically, what people are now worried about are not ten thousand Soviet warheads, but ten Iranian warheads or something like that. But if the United States had the ability to shoot down ten warheads, and some other country wished to blow up a nuclear weapon in the United States, then you don't send it by missile. You send it by ship or in the embassy pouch. This doesn't solve the problem. And what's more, it tells that country, "Build more than ten weapons."

Charlie Rose: My last point here is the notion that you have always believed there is life in another place because it would be an ultimate conceit of ours to think that we were the only place where there was enough intelligence to—
Carl Sagan: "Belief" is strong, but I would say it is such an important question. We have the ability to find out the answers, to send spacecraft to nearby worlds, use radio telescopes to see if anyone's sending us a message from a planet of another star. I'd be ashamed of my civilization if we had the tools to find out the answers and refused to look.

Charlie Rose: And if we didn't at least open ourselves to the idea that it's possible.
Carl Sagan: Absolutely.

A Slayer of Demons
Psychology Today / 1996

From *Psychology Today*, January/February 1996, pp. 30+.
Copyright © Sussex Publishers, Inc. Reprinted by permission.

PT: You've been most associated with issues of outer space. But you have turned very much to a world of inner space, the human mind.
CS: Well, the boundary between space and the Earth is purely arbitrary. And I'll probably always be interested in this planet—it's my favorite. I've written a number of books that have to do with the evolution of humans, human intelligence, human emotions. So it isn't a new departure for me to be concentrating on humans. Most of the people that I deal with are human. So I've had a lot of experience with that.

PT: Some of your best friends are humans. Your new book, *The Demon-Haunted World*, seems at times a litany of how the mind is fooled: by its own memory, by its senses, by shoddy reasoning. Is there intelligent life on Earth?
CS: Well, sure. But our intelligence is limited, and who would have expected otherwise? We're imperfect, and wisdom and prudence lie in understanding our imperfections. If we ignore our imperfections on the grounds that it's too depressing to concentrate on them, then we greatly limit our future options. On the other hand, if we know where our limitations are, not just in thinking but in emotional things, if we know about any hereditary predispositions we have towards ethnocentrism, xenophobia, dominance hierarchies, then we have a chance to moderate those tendencies. If we ignore any genetic predispositions in those directions, then we don't make any serious effort to ameliorate them and we're in much worse shape. This is one of those issues that every generation has to learn anew, because every generation has the same hereditary predispositions.

PT: But some of the issues you address in the book seem especially endemic to present times: UFOs, repressed memory. Are these kinds of things cropping up now more than before, as we approach the millennium?

CS: No. If you concentrate on the first few centuries of the Christian era, let's say, or the time of Mesmer in France, or almost any time in human history, you find just as many examples as from our present time. This is an endemic human characteristic—to be credulous, to believe what others tell us, to prefer what feels good to what's true.

PT: But until now, we've never been able to blow ourselves up. . . .
CS: Quite right. The dangers of not thinking clearly are much greater now than ever before. It's not that there's something new in our way of thinking, it's that credulous and confused thinking can be much more lethal in ways it was never before.

PT: You point to the statistical likelihood of people in power periodically showing up in the guise of a Stalin or a Hitler. Given this probability, and given nuclear proliferation, what are your feelings about the future?
CS: Well, it's a very serious issue. We are, fortunately, in a time when the United States and the former Soviet Union are divesting their nuclear arsenals. According to the present treaties, agreed to if not ratified, each side will go down to something like three thousand strategic weapons and delivery systems by the first decade of the twenty-first century, from ten times that number. So that's very good news. On the other hand, there are only about two thousand three hundred cities on the planet, so if each side gets three thousand weapons, that means that each side retains the ability to annihilate every city on Earth. That is certainly not comfortable news, because if you wait long enough you are *bound* to have a madman at the helm in one of these countries.

PT: Are you saying it's inevitable?
CS: If you look at the history of the world, such people regularly come to power. We may comfort ourselves in the United States that it hasn't happened to us, but even here I would say that a number of times in our recent history we've come close to having somebody dangerously incompetent or drunk or crazy in power in a time of crisis. Hitler and Stalin are reminders that the most advanced countries on earth can have such leaders.

PT: You spend a good deal of *The Demon-Haunted World* talking about, to use your term, scientific illiteracy. What do you think we should do? Clearly everything is going in the wrong direction.

CS: Well, the first thing I would say is that every generation has bemoaned the supposed lack of education of the next generation, and that goes back to some of the earliest Sumerian tablets that we have, from about five thousand years ago.

PT: With elders complaining about the youngsters of the time?
CS: Right: "They're not nearly as sharp as they were in my generation. They're not motivated. They don't do homework." So, there's always a danger of crotchety, elderly people comparing their generation with youngsters and concluding their generation was much harder working, more serious, had better values, better music, and so on.

Nevertheless, it's clear that there's a rampant dumbing down in progress in which not knowing things is considered a virtue and in which knowing things is considered a cause for embarrassment. I don't throw up my hands in despair. But I do try to indicate that it's a very serious problem that has no single point to face.

It isn't that if you were merely to increase the salaries of schoolteachers, you would solve the problem. The problem is endemic. It works at every level. It works in the culture of children themselves. It works in the federal, state, and local government. It works in the media. It works in the school boards and taxpayers with school bond issues. There's not just one point of attack. And it's very hard to imagine a serious change unless there's a change of behavior at many levels by many different people. That involves rethinking, it involves changes in values, it involves money—not out of cynicism, but out of understanding how the real world works. It's going to be very difficult to make this change unless, as happened with Sputnik, there's an apparent threat to national security that requires us to learn more science.

PT: We need a Sputnik-like explosion in public awareness to make us think, wake up.
CS: We do have the example of the late '50s and the early '60s. I don't know if that's the only thing that can make us do it. A sudden outbreak of wisdom maybe would be such a shock.

PT: I don't think we should count on that. Sputnik worked in part, I think, because people then had faith that science was going to cure our medical ills and solve the world's problems. People today don't have the same view of science as a panacea.

CS: As someone whose life was saved in the last six months by medical science, I certainly don't share the skepticism. The lives of almost everybody on Earth depend in the most intimate way on science and technology—to be unenthusiastic about science and technology is not just foolish, it's suicidal.

Without agricultural technology, for example, the Earth could support only tens of millions of people, instead of billions. That means that almost everyone on earth, 99 percent of us, owe the very fact that we're alive and haven't starved to death to the existence of technology.

PT: You just referred to your own intimations of mortality. Has that changed your outlook at all? You've recovered from something that could have been very serious.

CS: It *was* very serious. It's a bone marrow disease called myelodysplasia, which is invariably fatal if not treated. I had a bone marrow transplant at the Fred Hutchinson Cancer Research Center in Seattle. I was lucky that my only sibling, my sister, was a perfect match. It was lucky, but also I was the beneficiary of decades of experience that institution, and medical science in general, has had in bone marrow transplants. The age at which you can get a transplant is increasing every year. I think I'm the oldest person to get a transplant.

PT: Science saved your life.

CS: This is not the first time I almost died. This is my third time having to deal with intimations of mortality. And every time it's a character-building experience. You get a much clearer perspective on what's important and what isn't, the preciousness and beauty of life, and the importance of family and of trying to safeguard a future worthy of our children. I would recommend almost dying to everybody. I think it's really a good experience.

PT: Probably once is enough for most people. In part because science has done such a wonderful job of saving lives, we have a population crisis, at least in some people's eyes. Does that worry you?

CS: Oh yes, absolutely. But it's also clear how to resolve the problem. It involves complex social issues, and there are religious and nationalistic objections to dealing with the crisis. As with all crises, it will, if untreated, blow up in our face. The way to treat it is very threatening, since it is the billion poorest people who reproduce fastest, for simple reasons of survival. If you have children and no Social Security, there's a chance that some of your children might survive into your old age and take care of you. It's a simple calculation that the poorest people make, to have lots of children. So the first thing to do is to improve the self-sufficiency of the billion poorest people on the planet, which will lessen the charity of the major religions. It's not just good ethics, it's good in the most practical sense.

There also has to be a ready supply of safe, easy-to-use contraceptives. And the third key item is the political empowerment of women. There are societies in which the per capita income is high, but women are so oppressed that they cannot have a say in whether or not they have children. There are good reasons for helping the poorest people, and good reasons for empowering women, apart from the population crisis. But the population crisis makes it very clear that those should be prime goals.

PT: You're not just a scientist, you are also a celebrity. Because of that visibility you can be a salesman for certain issues if you care to.
CS: Since childhood, the most pleasurable occupation I could imagine was being a scientist. It had a romance to it that nothing else I know of even approached. And I've never lost that. My goal always was to be just a working scientist. It's true I studied some very exotic areas of science. I was interested in exploring other planets at a time when man had not even gotten outside the earth's atmosphere. So I actually have spent much of the last thirty-five years exploring the solar system, my childhood dream.

But, at the same time, I'm a citizen, a parent, a grandparent. I'm concerned about the future for all sorts of readily understandable mammalian reasons, and I would much rather work hard to make a better future, even if I fail, than to make no attempt.

PT: Do you spend half your time doing research and the other half doing soldier's duty as one of the world's most famous scientists?
CS: I don't try to budget my time from one to the other. They sort of naturally flow into one another. For example, I did my doctoral thesis on the Venus greenhouse effect, never imagining that the greenhouse effect would be a major global policy issue thirty years later.

There are several other cases—nuclear winter is one—in which the science and the public policy effortlessly flowed into each other. And the most natural thing in the world, if you find a science that you're to some degree expert in, is speaking out about a danger to the global civilization of the human species. If you won't, who's going to speak out? I just don't see it as two hermetically sealed compartments that you hop from one to the other. It often just flows in the most natural way.

I do have an opportunity that, unfortunately, others who are equally or more capable sometimes don't have, of communicating to the general public. And it's an opportunity that ought to be used carefully, not squandered. And used responsibly. But if I have opportunities to speak to the public, then certainly I'm not going to say no if I have something to speak for.

PT: Do you still have the same sense of wonder over science as you did twenty-five years ago?
CS: Last week, a planet seems to have been discovered around a nearby star called 51 Pegasus. And it's a planet very close to the star, much closer than Mercury is to our sun. But it's not a little rocky world like Mercury or Venus or the Earth. It's a giant world, presumably like Jupiter.

What is such a massive planet doing so close to that star? Does it have other terrestrial-type planets further out? Is that planet a gas giant the way Jupiter is, or is it a monster earthlike planet? And what does it say about the abundance of planetary systems elsewhere? Maybe they're all like that, and ours is anomalous. If that's true, what implications does that have for the origins of solar systems? I don't know. My wonder button got pushed hard when that discovery was announced. And it happens regularly. It certainly happens in my own research, such as in the laboratory work that we do on organic chemistry and the outer solar system, the origin of life on Earth. My wonder button is being pushed all the time.

PT: When you look at fellow scientists who are not, say, twenty-five or thirty anymore, do they still have the ability to wonder?
CS: Some do, some don't. Some lose it.

PT: What makes it go?
CS: One thing is a kind of Peter Principle. Good scientists are eventually offered opportunities to be administrators. That takes them away from science.

To be the department chairman, the president of a professional society, or a presidential science advisor, or whatever—those are all responsible and important positions, even ones that can aid the advancement of science. But not by you doing the science yourself. It's very hard to continue doing the science in some of those positions. They are very time consuming. So that's one danger.

Another thing is, the wonder is almost instinctive—you can see it in children—but the skepticism has to be learned. And you learn it sometimes by painful experience. You have experience with baloney, so your baloney-detection ability improves. If you never encounter baloney, then there you are, with all wonder and no skepticism.

So as time goes on there's a tendency to become more and more skeptical and to mistrust wonder. Very dangerous, because it's the balance between the two that's needed. So in a lot of scientists, the ratio of wonder to skepticism declines in time. That may be connected with the fact that in some fields—mathematics, physics, some others—the great discoveries are almost entirely made by youngsters.

PT: Was Einstein at the end of his life a man who had the capacity to wonder?
CS: No question about it, absolutely full of wonder.

PT: You've said that when you were growing up you didn't realize somebody could do science for a living. You envisioned being a salesman or something and doing science on weekends and evenings. It's all too rare that someone as young as you were at the time becomes so enthralled with science. Are we essentially killing off the wonder in children?
CS: Every kid starts out as a natural-born scientist, and then we beat it out of them. A few trickle through the system with their wonder and enthusiasm for science intact.

PT: Why did yours stay intact?
CS: The main thing was that my parents, who knew nothing about science, encouraged it. They never said, "All in all, wouldn't it be better to be a lawyer or a doctor?" I never once heard that from my parents. They said, "If you're passionate about that, we'll back you to the best of our ability." In school, while there were very few teachers who excited me about science, there was no systematic effort to discourage me.

So it wasn't that hard to maintain my interest. Science fiction sustained me in my earliest years. I got a keen sense of the excitement of science from science fiction.

PT: What is the dumbest thing you've ever done? I mean that affectionately.
CS: Oh, there are so many competing candidates. In fact, in this book I list some of the times where I've been dead wrong; in past books I've tended to stress the cases where I've been right, like the greenhouse effect. I suppose that's a natural human failing, but I've tried to make up for it a bit. Mistakes, wrong guesses, invalid conclusions are not disasters in science. In many cases they spur others to disprove or to check you out. And so it advances the field. The greatest scientists have made mistakes.

But one of the beauties of science is that it has built-in error-correcting machinery. Science, unlike many other human endeavors, reserves its highest rewards for those who disprove the contentions of its most revered leaders. Think, for example, of religion. How foreign that scientific point of view is from the religious idea, which so often is to uncritically accept whatever the founder of the religion said. It's not a tragedy that scientists make mistakes, and I certainly have made some in my time.

PT: Coming as you do from a hard-science background, how do you think psychology is doing as a field? A lot of the issues in your book are big areas in psychology.
CS: I'm not a psychologist. I don't have a comprehensive surveillance of the whole field, so all I can do is give you an offhand impression.

The thing I've been most appalled by is the sense of so many psychotherapists . . . that their job is to confirm their patients' delusions rather than help them find out what really has happened. It took a long time to convince myself that's what's happening, but it certainly *is* happening. I don't know whether it's more likely among social workers than Ph.D.s in psychology, or more likely among the Ph.D.s than the psychiatrists, who have medical training. But I do find it astonishing that anybody in psychology should be ignorant of the most elementary precepts of skeptical scientific scrutiny.

As someone who spent a lot of time reading Freud and his followers, I also am distressed by the absence of a systematic effort to demonstrate that psychoanalysis is more useful than going to your priest or rabbi. Or whether there is such a thing as repression. It's always very dangerous when the error-correcting machinery is not working and there aren't systematic attempts to disprove what the revered founder of your field maintains.

On the other hand, I see spectacular potential in imaging analysis of brain function. That is an amazing development, and you can see really major understandings of brain function coming out of that. Also tremendously exciting is the work on neurotransmitters, work on endorphins, and on the small brain proteins. Those are all tremendously exciting, and all of them, by the way, tend to support the idea that the mind is merely what the brain does. There's nothing else, there's no soul or psyche that's not made out of matter, that isn't a function of ten to the fourteenth synapses in the brain.

PT: As someone who has argued so eloquently about the role of evidence in making decisions, what is your reaction as a citizen and scientist to the O. J. trial?
CS: There are a lot of studies of juries that suggest that people make up their minds in the opening arguments, selectively remember the evidence that supports their initial judgment, then simply reject the contrary evidence, put it out of their heads. I suspect that did happen here.

The fault lies with prosecutors for relying on complex scientific and mathematical arguments without explaining it in a way the average person can understand. It was a failure to understand what is necessary in talking to the public about science. When we hear that the chance of this blood being someone other than O. J. Simpson's is one in one hundred billion, and there are only 5.5 billion people on the planet, and that is intended as a knock-out punch. . . . If somebody has no knowledge of elementary probability theory, the prosecution has an obligation to explain it step by step, from there being one chance in two when flipping coins, to highly improbable events.

Likewise, I think many jurors, many Americans anywhere, have little sense of what DNA is. They need some background on what DNA is, what are its unique characteristics, why it is different from person to person, the role it plays in determining heredity. There was none of that.

PT: Can that be accomplished in a trial?
CS: Sure. You do it in a very effective, humorous way with excellent visuals. It's pointless to bring to the public scientific and mathematical evidence if no one is going to understand what you're saying.

PT: You've done that as well as anyone.

CS: I'm often asked by colleagues what's the secret. Many scientists who are superb practitioners of their field claim that they're no good at explaining science, but I just don't believe that. I think there's only one secret. And that is, don't talk jargon. Don't talk as you would to colleagues. Instead, talk as you did to yourself at the time when you yourself didn't understand. You have to explain to people what's true in ordinary language, not technical terms. You have to respect the intelligence of your audience, but remember that they haven't had the advantage of the same technical education that you have.

PT: In looking for intelligence and originality in people, what earmarks do you use?

CS: I look for enthusiasm and wonder, but there's such a thing as too much. I look for someone who knows what he or she is talking about, because there's a tendency to repeat anything you've read without skeptical scrutiny of it. But in meeting people, it's rare that what I'm impressed by is their intelligence. There's much more likelihood that what I'm impressed by is their compassion, their optimism, their sense of humor—things of that sort I find much more compelling. There are very few people who don't have an impressive degree of intelligence, especially children. Society does very dangerous things in squashing that intelligence. It's a tragedy. You can see a kind of Darwinian competition of nations, and the ones that squash the intelligence of the citizenry in the long run are not going to do very well. The ones that learn to encourage curiosity and wonder and hard work are the ones that are going to make it.

PT: Are there insights to be gained from nonrational thought, religious thought?

CS: Certainly the insight that we're capable of nonrational thought is to be gained from nonrational thought. That is something very important. Every society—there are no exceptions—has some kind of religion. That tells us something important about human nature. It doesn't say that what the religion says is true. It says that there is a common need, that must be genetically based, that religions make an effort, successful or not, to deal with.

PT: A drive to find meaning or purpose?
CS: It's partly that, and also the need to have a code of ethics, because otherwise society is impossible. A sense of community, communion with nature, communion with your fellow human beings. A sense of ritual, music, art, poetry. Religion appeals on many different levels and serves many different needs. It would have to, to be so widespread.

PT: You have a young son. What are your biggest fears for the world he's inheriting?
CS: There are so many. I'm certainly worried about local and global environment. About overpopulation and violence. I'm worried about stupidity. I'm worried about consumerism, the focus on buying things that by any survival standard you don't need, but which American advertising culture promotes like mad.

PT: What gets you most excited for him?
CS: The inexhaustible benefits that emerge from science. I don't just mean agriculture and medicine, which have a large variety of practical benefits. The thing I like most about science is its room for managing the future. It's a tool for baloney detection. It's absolutely essential, not just for the technological products of science, but as a way of thinking. If that were more widely understood, we'd be a lot more secure in the future than we are now.

PT: Aldous Huxley wrote *Brave New World* in 1932. Have you thought of writing a book about the future, say, a century later?
CS: Prophesy is a lost art.

PT: He didn't write a prophesy, he just took information—
CS: Well, more than that. He was trying to give us a glimpse of a future society we should avoid. It was a cautionary tale. That was one, but there are so many. There are already possible dire futures; you could spend the rest of your life writing cautionary tales. Anyway, I have no plans to do so.

PT: You did write a novel a few years ago. What inspired you to write it?
CS: It's called *Contact*. It's being made into a motion picture starring Jodie Foster. It's the story of the receipt of a first bona fide radio message

from another civilization in space, and of the response here on Earth, which is very complex and diverse. I wrote it because it was an opportunity to get across scientific ideas to an audience different from that of *Scientific American*.

Also, it seemed fun to try to write fiction. And many people have asked me what I think the consequences of receiving such a message would be. I never could give in a few sentences what seemed to me an adequate answer.

PT: Are you hopeful that there is intelligent life elsewhere?
CS: My mind is certainly moot. Monitoring extraterrestrial radio waves is a chance, at relatively small cost, to try to answer one of the deepest questions ever posed. It's the importance of the quest, and the fact that we don't know enough to say in advance that it's fruitless, that motivates me. But I don't pretend to know that there are beings out there.

Talk of the Nation: Science Friday
Ira Flatow / 1996

Transcribed from *Talk of the Nation: Science Friday*, May 3, 1996. © NPR® 1996. Any unauthorized duplication is strictly prohibited. Reprinted by permission.

Flatow: This is *Talk of the Nation: Science Friday*, I am Ira Flatow. Astronomer Carl Sagan has spent about two decades—a good part of his career—trying to make science more understandable and relevant to be non-scientists. From his early days as a visible spokesman for the Viking-Mars lander to his Pulitzer Prize for *The Dragons of Eden* to his landmark TV series *Cosmos*, Dr. Sagan has tried to show that science is a tool for exploring the unknown—for rationally investigating and answering the mysteries of the world we live in. Carl Sagan, master communicator, joins me today to tell us why the scientific method is so important, so elegant, and so successful, and why people who believe in aliens, UFOs, and ESP abandon critical thinking when they buy into pseudoscientific happenings.... Now let me formally introduce my guest. What can you say about Carl Sagan that hasn't been said already? Professionally, the Duncan Professor of Astronomy and Space Sciences at Cornell University in Ithaca, he is the recipient of more awards and doctorates than you can shake a stick at, including a Pulitzer Prize and also the highest award given by the National Academy of Sciences. His most recent book is *The Demon-Haunted World: Science as a Candle in the Dark*, published by Random House, and he joins us from station KUOW in Seattle. Welcome to the program!
Sagan: Thank you, Ira. Thank you for that generous introduction.

Flatow: Oh, well, generous—it's the facts, man, just the facts! Were you shaking last night in Seattle [during the May 2 earthquake]?
Sagan: Yeah, it was very interesting! You know, your first impulse is to think that it's internal, that you're dizzy or sick or something, and then you find that everyone else is having the same symptoms and you figure that maybe it's something outside of your head and not inside of it.

Flatow: Well, you are recovering, are you not?
Sagan: As far as I can tell, I'm all better. I've been very lucky.

Flatow: Okay, that's good—it's good to see that whatever was ailing you is not attacking you anymore.
Sagan: Uh, right—it's gone away.

Flatow: Let's talk a little bit about a book. When an author decides to write a book, they usually feel very strongly about something and I think that of all the books that you've written, this latest, *The Demon-Haunted World*, really shows that you're angry about something here—about pseudoscience and the lack of rational thought. Would that be a correct assumption?
Sagan: Well, I'm certainly concerned. I don't know if angry is right, although maybe it is—maybe I am a little angry that we have such great tools, such powerful mental apparatus at our command which we tend to ignore. I mean, science more than a body of knowledge is a way of thinking, and its enormous success is due to accepting contentions only on the basis of evidence—and compelling evidence at that. It doesn't matter if it feels good; what matters is if it's true. And naturally there are people who want what feels good—I mean, that makes a lot of sense—and if discovering that we're not at the center of the universe, or that we're not the apple of God's eye, and so on, if there is no afterlife or evidence for an afterlife, those contentions rub a lot of people the wrong way and they'd rather not hear from science on such issues. They'd rather have their own fantasies which make them feel good.

Flatow: But are we going through an unusual period? I was tuning through the dials last night on television, and there are shows called *Sightings* and other things like it that are just popping up all over the place dealing with psychic phenomena, alien abduction, all the kinds of things that you talk about in your book. Are we going through an unusual period in history where there's a tremendous popularity in this that there has never been before?
Sagan: No, I don't think so. I think this way of looking at things—embracing pseudoscience and superstition and fundamentalist zealotry—has been with us humans for all of our history. It's not surprising that we should find that it's still around. But what is a little surprising is that science—which is so successful, which is responsible for our lives in most cases—is so poorly taught, is so poorly understood, and that the kind of skepticism that we would use in

purchasing a used car is in many cases not in evidence on ESP and crop circles and literal interpretation of what's written in the Bible and so on.

Flatow: Yet you point out in the book that you yourself had a terrible science education, and look what happened to you—you went on to become interested in science. Could that not happen to other people?
Sagan: I'm sure it could. Growing up in the thirties and forties, I didn't have a good science education. Although I had lots of science courses in middle school and high school, it wasn't until I got to college that I had real science by people who actually understood it and understood how to teach it, and that was such a breath of fresh air. Today we spend a lot of money and a lot of time on science education in the schools, but very often it is inadequately taught—you know, why is the basketball coach teaching chemistry? Why is the science all from the book and so little from the laboratory? Why are teachers nervous when bright kids ask penetrating questions? Why do the varsity basketball, baseball, and football players get spiffy jackets that are attractive to the opposite sex, but expert mathematicians and scientists and historians and others do not get spiffy jackets? Who made those decisions? Why are we doing things that way? And that's a kind of hint of the nature of the problem—it runs up and down our society. Almost every newspaper in the country has a daily astrology column. Most don't even have a weekly science column. When's the last time you saw science discussed on those dreary Sunday morning insider political programs? When's the last time a president of the United States made an intelligent remark on science? And so on.

Flatow: I think that's a good point. My own personal feeling about this is that people do want to talk about science, they are very interested in the unknown, they are very interested in where we came from and where we're going, and so the only places they get to see anything like that is in these new breeds of programs that are on the air.
Sagan: Absolutely.

Flatow: So now at least here's at least an opportunity to let their mind expand and watch them and think about something, even if it's pseudoscience.
Sagan: My experience is that all children have an intact sense of wonder. When I teach kindergarten or first grade, I have a room full of scientists at least as far as wonder is concerned. They're not up on the skepticism quotient

yet, but that's fine—that's something that can be taught to them. But by the time they get to high school, when I talk to seniors, twelfth graders, in high school, it's all gone—there are no followup questions, they're not listening to what their colleagues are saying, they're worried about how their questions will be received by their peers, their minds have been turned off, the sense of wonder is almost gone. Something dreadful happens to students between first and twelfth grades, and it's not just puberty—the interest in science that is there in first grade is beaten out of them by twelfth grade. And I think part of it is that there are adults who are nervous about being asked penetrating questions by young people, and so they give offputting answers. "Why is the moon round?" "Well, what did you expect it to be, square?" Instead of encouraging the child—it's a deep question, why is the moon round? It can get to the nature of gravitation, central forces, the strength of materials, there's so much in there if you wanted to pursue it. And likewise all those other wonderful questions that kids ask—why do we have toes, what's the birthday of the world, how deep could you dig a hole, and so on. Every one of those is an aperture to exciting children with their natural aptitude of interest in science, exciting that and encouraging them not necessarily to be professional scientists, but to be citizens who have a responsible role in dealing with science. We have a society based on science and technology, and at the same time we've arranged things so that almost nobody understands science and technology. That's a prescription for disaster as clear as anything.

Flatow: And they then look towards scientists, or people of science who explain to them how things work or what's wrong with society, and yet at the same time a lot of people—because scientists work for "the government" or are paid for by big universities—they're also distrustful of what they hear that scientists have to say, thinking that, that this is just another government coverup. A lot of the politics of the age we're living in has also filtered into the world of science, and this has been going on for years.

Sagan: Well, I think it should be said, Ira, that scientists—because science and technology are so powerful—scientists have provided instruments of destruction. It really is true that scientists are in some sense responsible for nuclear weapons, which could destroy the global civilization, maybe the species. Science has played a key role in the means by which the ozone layer is destroyed and by which global warming is happening, and so it is natural—especially if scientists are not in the media explaining what they're about—for people to mistrust scientists, and you see it in the Saturday morning cartoon mad scientists caricature, which is very prevalent.

Flatow: I also find, though, that sometimes scientists—and I don't mean this about you—but a lot of times scientists speak down to people; they are condescending. As a journalist I've had it happen to me many times but that's part of my business, but I also think that when people ask a question, especially about something that they don't understand, if it's something that's in the realm of the paranormal or they're truly questioning it, that scientists will say that's not an area we're going to spend any money to investigate—why do you ask such a silly question, we're not going to look into it, and just go away. Much the way probably their teachers talked to them. Wouldn't this affect the fact that people aren't looking for scientists to answer the topics that you answer in your book? I think this is a failing of scientists to answer these questions.
Sagan: I agree with you again, Ira. The general attitude of many scientists is that such questions are interesting and important but that the work that's been done shows that it's very unlikely there's anything to it. But that's very different from saying that asking about ESP is a question beneath contempt—no scientist should do that. You don't dismiss questions before you look into them, but only after you look into them. It's sort of the difference between prejudice and what I might call post-justice. Post-judice is perfectly okay, but prejudice is not. There has always been a fraction of the scientific community that not only dismisses such questions, but dismisses the whole idea of explaining what they're about to the public. In the sixth century BC, the Pythagoreans discovered that the square root of two was an irrational number—that is, could not be represented as the ratio of any two numbers, no matter how big. This information about the irrationality of the square root of two was promptly classified top secret, and there was a Pythagorean who made the mistake of explaining it to the public and when his ship went down and he drowned, Pythagoreans all over the Aegean nodded their heads saying, "You see, the gods have stepped in to prevent the popularization of science." No, I think scientists have an obligation—if nothing else, for selfish reasons—to explain what they're doing, to explain the joy and power of science. We live in a democracy, the people are supposed to have something to say about what the government does. Every day there are scientific issues being legislated on, and how can we instruct our elected representatives if we don't understand what the issues are? I mean, AIDS and cancer and superconducting

supercolliders and decaying infrastructure and should we send people to Mars, all of those questions, genetic engineering and many medical issues, all those questions involve science. We must understand those issues just for our own well-being. Then there are economic questions. There are industries that are fleeing American shores because Americans at the entry level are insufficiently educated in eighth-grade arithmetic or whatever it is to produce quality products. Then there's the fact that science in our time has been able to approach the deepest questions of origins—something that every human culture has been interested in and has spent some resources on. Where do we come from? Where does life come from? Where does our planet come from? Where does the whole universe come from? We actually have some preliminary answers to those questions, and you have to be made out of wood not to be interested a little bit in that. People are so grateful to learn some of the tentative answers to these questions. And then finally, that skeptical, questioning, don't-accept-what-authority-tells-you attitude of science is also nearly identical to the attitude of mind necessary for a functioning democracy. Science and democracy have very consonant values and approaches and I don't think you can have the one without the other.

Flatow: Let's go to Mike in Juno, Wisconsin.
Mike: Hi, Ira, thank you very much. It's a pleasure to finally get to talk to you, Mr. Sagan. I met you briefly at the space science building in Ithaca some years back and I gotta say I can't agree with you more on there not being enough of this exposure in the media for people like myself and a few others and in addition I don't think it goes far enough. But what I wanted to ask you years ago when I met you and I'd like to ask you now is a very personal question. What is your real belief in the spiritual beginnings of all of this? In other words, I've read many places that many scientists such as yourself are either agnostic or atheistic in their beliefs about the initial beginnings of the universe, and secondly my brother-in-law is a highly classified person that works for the Air Force—the question revolves around what your belief is on UFOs. He's told me some things that I can't repeat about the Roswell incident, perhaps you could give us your thoughts on that.
Sagan: Well, in 1947 near Roswell, New Mexico, stuff came down on a ranch, and the stuff was then picked up by Air Force personnel. People were apparently told to keep quiet about it, and over the years the story has emerged that these were parts of a crashed alien spacecraft and that little alien bodies

were shipped to an air force base in Ohio and that they're still languishing in freezers with their perfect teeth. The actual fact seems to be that as the Air Force announced very belatedly just a couple of years ago that this was a balloon at tropopause altitude—just where the stratosphere begins—with acoustic instruments designed to detect Soviet nuclear weapon explosions from halfway around the world. There is an acoustic channel at the tropopause by which you might be able to hear such explosions. And this was a matter of the highest concern for the security of the United States and was properly classified. Newspaper photographs of the time show flimsy polyethylene-like material and balsa wood, hardly consistent with the spacecraft of an alien advanced civilization but perfectly consistent with balloons. I think that there are two museums of UFOs in Roswell, New Mexico. You can make money, you can get your name in the paper, you can have a break from the humdrum day by inventing stories about Roswell, New Mexico. I don't know what your brother has told you but I would treat it with a real grain of salt.

Flatow: Does that mean you don't believe in UFOs?
Sagan: Well, what do we mean by believe? Which, in fact, takes us to the first question. What do we mean by believe? If the evidence is compelling, then we believe. If the evidence is not compelling, then we don't believe—we withhold judgment. And UFO merely is an abbreviation for unidentified flying objects. If we see something in the sky and we don't know what it is, to my mind that's it—we don't know what it is. It does not automatically follow that it's spacecraft from somewhere else. The vast majority of UFO reports have quite prosaic explanations. People seeing natural phenomena in the sky with which they're unfamiliar, including cases of astronomers doing that, sometimes conscious hoaxes, sometimes people who hallucinate—and 25 percent of all people hallucinate—so there are many other explanations. And only if you've been able to eliminate all of those explanations would you give serious consideration to the possibility that we're being visited. Nobody's more interested than me in the possibility of the existence of extraterrestrial life. I've been involved in sending spacecrafts to other planets to look for it, I've been involved in using large radio telescopes to listen for signals from civilizations from planets of other stars. It would save me so much effort if the aliens were here—even if they are short, dour, and sexually obsessed, as the alien abductees, so-called, claim.
[laughter]

Flatow: And before we lose time this half hour, part one of his question—about your personal beliefs.

Sagan: Okay, well, I treat the existence of God and, perhaps, creation of the universe in exactly the same way. What is the evidence? Now, the word God is used to cover a wide variety of very different ideas, ranging maybe from the idea of an outsized light-skinned male with a long white beard who sits in a throne in the sky and tallies the fall of every sparrow—for which there is no evidence, none at all—to the view of Einstein, of Spinoza, which is essentially that God is the sum total of the laws of nature. And since there are laws of nature, and since remarkably the same laws hold throughout this magnificent and vast universe, if that's what you mean by God, then of course there's a God. So everything depends on the definition of God. One last point: You ask about the origin of the universe, but that's assuming the issue in question, namely that there was an origin of the universe. And in some cosmological models, the universe is infinitely old, therefore uncreated, therefore there is nothing for a creator to do. So I think these are very deep and difficult issues in which both theologians and scientists ought to bear in mind their own limitations before the difficulties of these issues.

Flatow: Carl, reading [*The Demon-Haunted World*] a person could come away saying "Well, I guess that Carl Sagan believes that scientists are the only ones with the right answers, that there is only one answer—science—and that only scientists then know what the right answers are." Would that be a correct assumption?

Sagan: Well, depends what you mean by science. If by science you mean that you bear in mind human fallibility, and you treat claims to knowledge skeptically, then I would agree science is the only way to go. But that's a very broad definition. Everybody, as I was saying before, who buys a used car would then be a scientist. Science is only a Latin word that means knowledge, and we shouldn't imagine that it's something very erudite and arcane. I just think the key point of science is criticism, debate, open inquiry, the willingness to systematize knowledge, to withhold belief until the evidence is compelling, and to listen seriously to criticism.

Flatow: Well, I find that is a criticism of many scientists that, because they are people and have human foibles, they are close-minded and will not listen to new ideas and are very much just the kind of people you say not to be. I'm sure you find some of those people also.

Sagan: I do indeed. Scientists are human beings—we all knew that. And they have the foibles of human beings. But it is the mutual process of science, all the scientists working together, which makes it work so well. That is, yes, Scientist A may propose his theory and be irredeemably attached to it, and Scientist B may criticize the theory because B is jealous of A or B is ambitious or whatever you want to imagine, but A and B together get a debate going that other scientists then enter, and this is the aperture to understanding, finding out what is really true. Science is an enterprise that gives its highest rewards to those who disprove the views of its most revered figures. Religion is exactly the opposite. Religion doesn't want any criticism of its most revered figures. Science makes progress—after all, the person who disproved the ultimate validity of the views on mechanics and gravitation of Isaac Newton was Albert Einstein. Newton a tremendously revered figure, and Einstein revered in part because he proved Newton wrong, or wrong beyond a certain range of parameters. And that's what I mean—it's the collective enterprise of science which has the virtues I described. The individual scientists are, of course, flawed as we all are.

Flatow: And scientists can be very cruel to one another when they disagree. I'm speaking specifically, for example, like the last seven years of cold fusion, where scientists have cut off or blacklisted just about any scientist who is secretly working in his basement because he believes there's something going on—maybe fusion is the wrong word for it—but there's something going on in those little jars, and they continue to work in secret, or go to Europe, or get patents in Europe that they can't get in this country. [They've] been basically cut off from any discourse with their colleagues.

Sagan: But the world community of scientists is pursuing it, and German and Japanese companies are handsomely funding this, and if there's anything to it that will come out also. I don't think it's a matter of cruelty. I think that claims were made without sufficient evidence that were then disproved, and that left a bad taste. Some scientists went too far and said anything remotely like this is nonsense or fraud . . . but the idea that it's fusion, that neutrons or gamma rays are produced, that more energy comes out than you put in, that's very dubious.

Flatow: But you will agree that there's something there that scientists—credible scientists—are looking at.
Sagan: Yes, there is something, but the question is: what is it? And many people think it's in the realm of chemistry, and not of nuclear physics.

Flatow: How much of our culture influences the direction scientists take? We can go from cold fusion to something else—if a culture, maybe it's an eastern culture that believes in body transport or something, that channeling is possible, wouldn't that culture say, "Hey, let's take that direction and use our scientific technique to prove or disprove it," whereas in the western culture you'll go in a totally different direction.
Sagan: Sure, and that's very healthy. But any claim of body transference—I'm not even sure what it is, but whatever you have in mind—any claim that it happens has to satisfy committed skeptics. It has to convince them. It has to be compelling evidence. So sure, which topic scientists pursue is a very complex process involving cultural attitudes and personal ambition and wanting to pick a topic that's soluble in a lifetime and many other issues, but if you're going to get anywhere with it, you have to provide the compelling evidence which is at the heart of science.

Flatow: Even in physics research going on here, you mention in published interviews that you couldn't convince a court of law about UFOs, for example, much less a court of scientists. But let's say hypothetically that if I decided I was going to compare two places of research, I'm willing to bet you that using your own idea of convincing a court of law, if I asked a jury to believe in some theories of superstring physics, for example, where they ask us to believe that something could exist in twenty-four dimensions that then collapse down to four dimensions or ten dimensions, that a jury would have more trouble believing the validity of such a wacky idea than the testimony of the thousands of people who have seen or supposedly have seen UFOs.
Sagan: May be, but string theory is by no means accepted and there are very distinguished skeptics, like Steven Weinberg at the University of Texas at Austin, and I think the most you can say about string theory is that it's promising. But of course there are things that are technically too difficult for unselected juries, as there are in matters of tort law and so on, there are cases where juries are chosen to be expert, or where a judge is chosen rather than

a jury, for precisely that reason, but that doesn't mean that the issues are either clearer or less clear because they're complex.

Flatow: Let's go to Pete in Seattle. Hi, Pete!

Pete: Hi. Thanks for taking my call. Mr. Sagan, I'm a big fan of yours and I think I'm actually a few miles from you here in town. What I wanted to say was that I think a lack of critical thinking skills is practically almost the root of all evil. Successful critical thinking goes way beyond scientific topics. It goes into every part of our life—social behavior, economics, morality, ethics, and all that kind of thing. I'm wondering what we can do to try and do a little more organized teaching of critical thinking skills, maybe in the schools, to everyone really, but I think it would going a long ways towards—I hate to use the phrase "raising the lowest common denominator,"—but I think that's kind of what it would be towards, all of us kind of getting together and choosing to evolve and develop our minds kind of as a goal. It seems to be put on hold right now.
Sagan: Yes, well, part of the problem is that you start teaching young people critical thinking and they'll start criticizing their political institutions—

Pete: That's good!
Sagan: —and religious institutions—

Pete: That's all right!
Sagan: —yeah, but then the people in power say, "Oh my God, what are we doing?"

Pete: Well, I'm thinking about trying to get people in power as part of the students in this process, you know.
Sagan: Yeah, but I think people in power have a vested interest to oppose critical thinking.

Pete: Yeah, they sure do.
Sagan: If we don't improve our understanding of critical thinking and develop it as kind of second nature, then we're just suckers ready to be taken by the next charlatan who ambles along, and there are lots of charlatans, there are lots of ways of gaining power and money by deceiving people who

are not skilled in critical thinking. So what you suggest is absolutely essential, but getting it done is very difficult since there are so many institutional impediments.

Flatow: I think people are used to—because this is the way that a lot of life works, ever since you were a child—are used to dealing with black and white issues. There's either a wrong or a right, a winner or a loser, a ball game in between, and then there's a winner and a loser—and they're not used to accepting that there are gray areas and possibly many answers to a question.
Sagan: That's right. There's an intolerance for ambiguity: "Don't give me the alternatives, just tell me what's right." But in many cases, because humans are imperfect, we don't know what's right and it's essential to give the various views. And what's more, if we're forced to confront views opposing our own, then we can test our own views and see if they stand up, and if they don't, why would we want to hold on to them?

Flatow: Thanks for calling, Pete. Dr. Sagan, what happened to the scientist as a social and political activist? Linus Pauling is not around anymore; Paul Ehrlich is a little more active these days than he has been in the past; you yourself were out there, still are politically active, or at least socially active, speaking out when you think things should be spoken about, but you don't see the role models of scientists that people think of and can use and can rally around, even if it's just purely on scientific issues.
Sagan: Well, I think there are many such cases but they don't get much publicity. For example, there's Ted Postal at MIT, former civilian scientist with the Joint Chiefs of Staff, who examined the Patriot missiles . . . and showed that the evidence seems to show that not a single Iraqi scud was shot down, despite the claim of the manufacturer that the destruction rate of Iraqi scuds was enormous. Now this is the effort of a single scientist who happened to have some expertise in military hardware who, out of conscience, decided that he was going to blow the whistle on these guys. And there are lots of whistleblowers—the tobacco industry is beginning to show scientists with some conscience—you see lots of it. But the issues that, say, Linus Pauling was connected with, were critical issues of public health and life and death, and likewise Paul Ehrlich, which necessarily gathers more public interest than questions of whether scud missiles were shot down or not in the Iraqi Gulf War.

Flatow: Let's go to Larry in Brooklyn. Hi, Larry.

Larry: Yes. Hi, doctor! I agree entirely—to show a little of my thunder, this whole rise in the belief of mystical phenomena that you see on TV now, the Psychic Hotlines, unbelievable. In Wiemar Germany, and now even in post–Soviet Russia, there's a tremendous rise in this. Do you see a tie-in between totalitarianism/fascism and inability to think critically on social and political issues?

Sagan: Absolutely. After all, the dictators don't want people to be critically assessing what they say—they merely want their citizens to accept what they say, and believe and do. You have statements by Hitler, for example, that were very clear on this in which he says that science is merely a convention, the truth is merely a convention, and that in his regime, people will adopt a different convention, the convention of will, whatever the Fuhrer wills is what people will consider true, and evidence has nothing to do with it. I think the trend, today, towards thinking that science is just another belief system, no more valid than any belief system, has a distinct totalitarian aroma about it.

Larry: It's scary. And just one brief question about teaching youth science—it's a question about reification and animation. In other words, commonly when we talk to kids, we'll say two magnets attract each other, and/or the electron wants to do this or wants to do that, and imbue it with anthropomorphized [crosstalk] and deal with human attributes, or the question what is a field, and we sprinkle iron filings when it's just a graphic representation. There is no electron per se—we really can't sense it, we can't smell it—we only know it by how it acts upon other things, meters, devices, measuring devices. It's a hypothetical construct, and yet in common language we say electron or magnetic field. And to get this idea of graphic representation as being just that, confusing the road with the map to use a logician's terminology, how would you deal with this at the level of elementary and high school? It's hard.

Sagan: I quite agree. I don't think there's anything wrong with talking about an electron repelling another electron, or the Earth attracting an asteroid, and so on. They are convenient ways to think. But if we suddenly think that the Earth has a mind and is sexually enamored of the asteroid, and that's what we mean by being attracted to, then that becomes something quite different.

Larry: But even with professional scientists . . . they put the iron filings down and you can see them, and that's "the field."
Sagan: That's a tracer of the field. I think the Faraday Maxwell theory of fields has been tremendously productive, and explains the world very well, and that's all we ask of it.

Larry: But we can go no further and say, "I've seen an electron." Nobody has seen it.
Sagan: But you see, when you pick up a book, let's say, you have the sense that here's something solid and real and I don't have to deduce anything, but in fact you're not touching the book. The electric field of your fingers is interacting with the electric field of the book and there is in fact no physical contact, but that seems so contrary to common sense that we don't teach it. I think everything we know has this abstraction from reality, but all we ask of science is that it predictably explain the reality that we see. That it is ultimately true may be beyond what humans can do, if that phrase has any meaning at all.

Flatow: You know, some of the greatest paradigm shifts—that's a great phrase—some of the greatest changes of our society have come from popularizing events via mass media. Three Mile Island may have been bad news for the nuclear industry, but *The China Syndrome* movie was really, probably, worse than the accident itself. Popular movies and culture have a way of changing people's views. *E.T.: Extra Terrestrial* started having people talking about life in outer space. You've written lots of books about science. Do you believe that television—and the *Cosmos* series being the most spectacular and popular series of its kind—do you still believe that the popular medium is the way, no matter how many issues of *Scientific American* are sold or whatever, that this is the way to influence people's views about science?
Sagan: I think that television is a tremendously useful and powerful and underused medium for exciting people about science, for eliciting their sense of wonder, and for teaching some science fact, but mainly about getting people excited so they will go off and teach themselves or take courses or something of that sort. *Cosmos* we never imagined would be as successful as it was; it's been seen by more than half a billion people in more than sixty countries worldwide, and it's still being seen, and I still get letters and I'm stopped on the street by people who say that it changed their lives—women, especially,

say they were taught that science wasn't for them, that they were too stupid for science, then *Cosmos* got them excited about science and then they went back and now they're an oceanographer or a microbiologist or whatever it is—there is a tendency to discourage people from science, especially in junior high and high school, who are well-fitted for science. We have a kind of fear of science and part of the reason is that science is able to show what constitutes a wrong, and unlike some other fields where no matter what you say might be right, here in science you can actually make a mistake and have to defend your view to other people who can actually draw upon facts to disprove it. So it makes some people nervous, the people who want the world to conform to their wishes rather than to the universe's own internal reality.

Flatow: You were telling me that you were working on a movie. Are you at liberty to talk about that at all?
Sagan: I can talk about it a little bit.

Flatow: A new movie based on *Contact*, your book?
Sagan: Based on my novel *Contact*, about first contact with extraterrestrials based on receipt of a radio message. It's a Warner Bros. movie; it's starring Jodie Foster, and it's in production. Primary photography will begin sometime later this year, and it's unclear when it should be in the theaters but late '97 at the earliest.

Flatow: Can you teach science through this movie?
Sagan: I'm certainly working hard to get across some of the wonder and some of the method of science, and I think we're going to have some of that. You know, the big screen is an amazing tool for teaching the wonders of astronomy, especially—I can't wait to see how some of the ideas that we're having are going to materialize on the big screen.

Flatow: You know, Kubrick's film *2001: A Space Odyssey* I thought was a milestone in teaching science, the things that went on in there that were absolutely—about gravity and space travel and things like that.
Sagan: Right. It's amazing how *2001* stands up today. It doesn't look the least bit dated, whereas *2010*, its successor with the non–Stanley Kubrick director, looked obsolescent when it came out, and today is just terribly dated. So you

can do these things well and you can do them poorly, and we're hoping with *Contact* to do it well.

Flatow: So you'd be happy to have the *2001* success that movie had? That level?
Sagan: I'd be happy to come anywhere within shouting distance of *2001*. That was an extraordinary movie.

Flatow: On a serious side, though, this is a good way to reach the public and to teach science?
Sagan: Movies and television can do amazing things in teaching at least some science, but mainly in making science accessible and convincing people that they don't have to worry that they're too stupid to understand it or that it's stuff that only nerds and geeks are interested in. I think everybody is interested in many of the issues of science, and it's just a question of getting it to them in an accessible way.

Flatow: Carl, stay well!
Sagan: Thank you so much, Ira! Pleasure talking to you.

Flatow: Thanks again for coming on the program. Carl Sagan, of course, is a professor of astronomy and space sciences at Cornell University in Ithaca, and author of an extremely good book *The Demon-Haunted World: Science as a Candle in the Dark*, published by Random House.

The Charlie Rose Show
Charlie Rose / 1996

From *The Charlie Rose Show*, May 27, 1996, transcript #1647. Reprinted by permission.

Charlie Rose: Carl Sagan is one of the preeminent astronomers of our time. He is known for bringing the heavens to our living rooms with his PBS series *Cosmos*. His latest work is *The Demon-Haunted World: Science as a Candle in the Dark*. It explores the country's growing fascination with pseudo-science—astrology, faith healers, the supernatural and the like—all superstitions that he says threaten to undermine true science. I am pleased to have him here and I also take note of the fact that he is a David Duncan professor of astronomy and space sciences and director of the Laboratory for Planetary Studies at Cornell University, distinguished visiting scientist of the Jet Propulsion Laboratory, California Institute of Technology and co-planner and president of the Planetary Society, the largest space interest group in the world, and a former Pulitzer Prize winner. Welcome back today.
Carl Sagan: Thank you. It's great to see you.

Charlie Rose: Listen to this. I hate to read too much, but this is—it's almost like they've been reading your book. This is from the *New York Times* for Friday, May 24. "Americans flaunt science, a study finds. Less than half of all American adults understand that the Earth orbits the sun yearly, according to a basic science survey. Nevertheless, there's enthusiasm for research except in some fields like genetic engineering and nuclear power that are viewed with suspicion. Only about 25 percent of American adults get passing grades in a National Science Foundation Survey of what people know about basic science and economics." I mean, this is singing your song, isn't it?
Carl Sagan: Well, it's certainly what I'm talking about in *The Demon-Haunted World*. My feeling, Charlie, is that it's not pseudo-science and superstition and New Age, so-called, beliefs and fundamentalist zealotry are something new. They've been with us for as long as we've been human.

But we live in an age based on science and technology with formidable technological powers.

Charlie Rose: Science and technology are propelling us forward at accelerating rates.
Carl Sagan: That's right. And if we don't understand it—and by "we" I mean the general public—if it's something that, "Oh, I'm not good at that. I don't know anything about it," then who is making all the decisions about science and technology that are going to determine what kind of future our children live in? Just some members of Congress? But there's no more than a handful of members of Congress with any background in science at all. And the Republican Congress has just abolished its own office of technology assessment, the organization that gave them bipartisan, competent advice on science and technology. They say "We don't want to know. Don't tell us about science and technology."

Charlie Rose: Surprising, because Gingrich is genuinely interested, I think—
Carl Sagan: He is. No question.

Charlie Rose: —out of his own intellectual curiosity. Does the president still have a science adviser at the White House?
Carl Sagan: He does. He does—John Gibbons. And the vice president is scientifically literate.

Charlie Rose: He's well known for being a science maven. I mean, you blast them all—creationists, Christian Scientists who you say would rather allow their children to suffer than give them insulin or antibiotics. Astrologers come in for particular scorn on your part.
Carl Sagan: Well, I wouldn't say scorn, just derision.

Charlie Rose: A more generous version of scorn. But what's the danger of all this?
Carl Sagan: There's two kinds of dangers. One is what I just talked about, that we've arranged a society based on science and technology in which nobody understands anything about science and technology, and this combustible mixture of ignorance and power, sooner or later, is going to

blow up in our faces. I mean, who is running the science and technology in a democracy, if the people don't know anything about it?

And the second reason that I'm worried about this is that science is more than a body of knowledge. It's a way of thinking, a way of skeptically interrogating the universe with a fine understanding of human fallibility. If we are not able to ask skeptical questions, to interrogate those who tell us that something is true, to be skeptical of those in authority, then we're up for grabs for the next charlatan, political or religious, who comes ambling along.

It's a thing that Jefferson laid great stress on. It wasn't enough, he said, to enshrine some rights in a Constitution or a Bill of Rights. The people had to be educated and they had to practice their skepticism and their education. Otherwise we don't run the government: the government runs us.

Charlie Rose: Jefferson was amazing in his devotion to science.
Carl Sagan: Absolutely.

Charlie Rose: We think of Jefferson as this man who was literate and who was a passionate articulator of freedom, but if you go to Monticello, what you appreciate is he was at heart a scientist, a botanist, an architect, geologist. As we know from Stephen Ambrose, Jefferson wanted Meriwether Lewis to go out and do experimentations and explore and be skeptical and find answers to passages and explore the West.
Carl Sagan: Exactly right. And there was also an economic grail there if the northwest passage was found. Jefferson said that he was at heart a scientist, that he would have loved to have been a scientist. But there were certain events happening in America that called to him, and so he devoted his life to that kind of politics.

Charlie Rose: A revolution—
Carl Sagan: Indeed. So that generations later people could be scientists.

Charlie Rose: You have made the point: "When's the last time we had a president who made a speech about science." It is this notion that science is not of great interest to us in some sense, that somehow we don't want to learn.
Carl Sagan: You see, people read stock market quotations and financial pages. Look how complex that is.

Charlie Rose: Because they know the direct connection to their own—
Carl Sagan: There's a motivation. But they're capable of it—large numbers of people. People are able to look at sports statistics. Look how many people can do that. Understanding science is not more difficult than that. It does not involve greater intellectual activity. But the thing about science is, first of all, it's after the way the universe really is and not what makes us feel good. And a lot of the competing doctrine are after what feels good and not what's true.

Charlie Rose: OK. I'm not sure you'll go this far with me, but there's a lot of that that is about feeling good and there's a lot of that that's about hocus pocus. But at the same time, there are millions of people who understand science does not prove religion because religion is faith-based. Therefore, you should not deny the value of it because it is faith-based and not science-based.
Carl Sagan: But let's look a little more deeply into that. What is faith? It is belief in the absence of evidence. Now, I don't propose to tell anybody what to believe, but for me, believing when there's no compelling evidence is a mistake. The idea is to withhold belief until there is compelling evidence. And if the universe does not comply with our predisposition, OK, then we have the wrenching obligation to accommodate to the way the universe really is.

Charlie Rose: So you step forward to say, "I deny all religion because I can't see it proved scientifically"?
Carl Sagan: No, no, no.

Charlie Rose: You see the value of religious experience and the value of reaching for a higher experiences?
Carl Sagan: Religion deals with history, with poetry, with great literature, with ethics, with morals, including the morality of treating compassionately the least fortunate among us. All of these are things that I endorse wholeheartedly. Where religion gets into trouble is in those cases that it pretends to know something about science. The science in the Bible, for example, was acquired by the Jews from the Babylonians during the Babylonian captivity of 600 BC. That was the best science on the planet then. But we've learned something since then. Roman Catholicism, Reform Judaism, most of the mainstream Protestant denominations have no difficulty with the idea that humans have evolved from other creatures, that the

Earth is 4.6 billion years old, the Big Bang. They don't have any trouble with that. The trouble comes with people who are Biblical literalists who believe that the Bible is dictated by the Creator of the Universe to an unerring stenographer and has no metaphor or allegory in it.

Charlie Rose: And from there, they make their political and economic choices, and social choices.
Carl Sagan: And scientific.

Charlie Rose: And scientific choices. And that's part of your problem with that idea.
Carl Sagan: Exactly.

Charlie Rose: It is that—because for the wrong reasons, we make the wrong choices about science.
Carl Sagan: That's right. So who is more humble? The scientist who looks at the universe with an open mind and accepts whatever the universe has to teach us, or somebody who says, "Everything in this book must be considered the literal truth and never mind the fallibility of all the human beings involved in the writing of this book."

Charlie Rose: OK. I mean, I accept that, but the argument that would be made by many is that whether a specific scientific act took place as described by some Biblical writer is not at the heart of the religious faith and the religious experience.
Carl Sagan: Some people agree with you and some people don't. Some people think that every jot and tittle in the Bible is essential. You throw one thing away to allegory or metaphor; then it's up to everybody to make their own decisions.

Charlie Rose: A lot of this has to do with science in the United States? Are we different than other nations?
Carl Sagan: No. Absolutely not. You can see this worldwide. In India, there's a madness about astrology; in Britain, it's ghosts; in Germany, it's rays coming up from the Earth that can only be detected by dousers. Every country has its own specialties. We seem to be fascinated by UFOs right now.

Charlie Rose: Before you leave UFOs, tell me about you and Professor Mack.
Carl Sagan: John Mack is a professor of psychiatry at Harvard whom I've known for many years. We were arrested together at the Nevada Nuclear Test Site protesting U.S. testing in the face of a Soviet moratorium on testing. And many years ago, he asked me, "What is there in this UFO business? Is there anything to it?" And I said, "Absolutely nothing, except, of course, for a psychiatrist." He is a psychiatrist.

Well, he looked into it and decided that there was so much emotional energy in the reports of people who claimed to be abducted that it couldn't possibly be some psychological aberration, that it had to be true. He believed his patients. I do not believe his patients. Many of these stories are about waking up from a deep sleep and finding your bed surrounded by three or four short, gray and sexually-obsessed beings who then take you to their spaceship after they slither through your wall, and perform a variety of objectionable, sexual experiments on you.

Charlie Rose: But here we have Dr. Carl Sagan, astronomer, versus Dr. John Mack, M.D.
Carl Sagan: No question.

Charlie Rose: So what's the problem?
Carl Sagan: How can scientists disagree?

Charlie Rose: He's a scientist—he's a scientist. Well, no. I'm asking how could—I mean what do you think of this man coming to these conclusions?
Carl Sagan: I think he is not using the scientific method in approaching his issue.

Charlie Rose: And were you constantly—I mean, I assume you come at him with both barrels in conversations.
Carl Sagan: And in *The Demon-Haunted World*.

Charlie Rose: And he says?
Carl Sagan: He says I don't appreciate the emotional force of these reports. But many people awaken from a nightmare with profound emotional force. That doesn't mean that the nightmare is true; it means something went on inside our heads.

Charlie Rose: You were making a point before I jumped the gun.
Carl Sagan: What I wanted to say is going back to the question of adequate evidence on something that's emotionally really pulling you. I lost both my parents about twelve or fifteen years ago and I had a great relationship with them. I really miss them. I would love to believe that their spirits were around somewhere. And I'd give almost anything to spend five minutes a year with them.

Charlie Rose: Do you hear their voices ever?
Carl Sagan: Sometimes. About six or eight times since their death I've heard it.

Charlie Rose: Carl—
Carl Sagan: Just in the voice of my father or my mother. Now, I don't think that means that they're in the next room. I think it means that I've had an auditory hallucination. I was with them so long, I heard their voices so often. Why shouldn't I be able to make a vivid recollection of them?

Charlie Rose: Here's what's interesting about this for me—I mean, you won't see this, but I'll throw it at you anyway. You convinced me a long time ago that it was arrogant for me or for anyone else to believe that there wasn't some life outside of our—
Carl Sagan: To exclude the possibility.

Charlie Rose: To exclude the possibility was an arrogance of intellect that we should not assume. You couldn't prove it, you didn't know it was there, but the arrogance—
Carl Sagan: We don't know if it's there; we don't know if it's not there. Let's look.

Charlie Rose: And if you take that, why can't you say, "There's a lot we don't know. There's a lot of power there that we don't know."
Carl Sagan: I say it. It's what I believe, but that doesn't mean that every fraudulent claim has to be accepted. We demand the most rigorous standards of evidence, especially on what's important to us. So if some guy comes up to me, a channeler or a medium, and says, "I can put you in touch with your parents," well, because I want so terribly to believe that, I know I have to

reach in for added reserves of skepticism because I'm likely to be fooled, and much more minor, to have my money taken.

Charlie Rose: Well, is it J. Z. Knight—
Carl Sagan: Yeah, exactly. She has a guy named Ramtha who's ten thousand years old or something.

Charlie Rose: Thirty-five.
Carl Sagan: Thirty-five, yeah. And he tells you lots of things but nothing about what life was like thirty-five thousand years ago.

Charlie Rose: Shirley Maclaine believes.
Carl Sagan: Shirley Maclaine believes that Ramtha was her brother.

Charlie Rose: Things like the Loch Ness monster and all of that. Is it all faked?
Carl Sagan: The most famous photograph has now been shown to be a fake, but could there be a unknown mammal or even reptile of large dimension swimming in a Scottish lake? Sure there could. That we don't know about? Sure there could. Who says no? But the evidence does not support it, does not demonstrate it. So do we say, "Oh, ridiculous."? No, we don't do that. We say "unproved," which is a Scottish verdict.

Charlie Rose: Some reviewers differ with your conclusions on this point—that you seem to say it's growing, this kind of pseudo science, and—
Carl Sagan: No. Sorry to interrupt. I don't—this is part of being human. Humans have had this way of magical thinking through all of our history. The problem is that today the technology has reached formidable, maybe even awesome, proportions, and so the dangers of thinking this way are larger. Not that this is a new kind of thinking.

Charlie Rose: You are living with myelodysplasia.
Carl Sagan: Or I have been.

Charlie Rose: You have been. It's in remission. Are you—
Carl Sagan: Well, you know, with diseases of this sort and in all cancers—

Charlie Rose: Cancer of the bone marrow?
Carl Sagan: Myelodysplasia is not exactly cancer of the bone marrow, but if untreated, it inevitably leads to leukemia. And the trouble with all these diseases is that you never know that you've got every last cell. You can only detect down to a certain level. But down to the level that anybody can detect in terms of how I feel and my stamina and all that, it seems to be gone. I'm very lucky.

Charlie Rose: Because you had a sister who enabled you to have a bone marrow transplant.
Carl Sagan: That's one. And also the enormous advances in medical science in just the last few years. If I had had this thing five or ten years ago, I would be dead, sure as shooting. And then, finally, the love and support of my family. All of those played a central role.

Charlie Rose: So you're optimistic?
Carl Sagan: I'm very optimistic, or at least very hopeful.

Charlie Rose: And just share with us, because of your sense of language and your sense of understanding and being reflective and introspective, what does—what do you think about and what does it do for you to—
Carl Sagan: I didn't have any near-death experiences; I didn't have a religious conversion, but—

Charlie Rose: You thought about what it would be like to die.
Carl Sagan: Certainly. And what it would be like for my family. I didn't much think about what it would be like for me, because I don't think it's likely there's anything that you think about after you're dead.

Charlie Rose: That's it, huh?
Carl Sagan: Yeah, a long, dreamless sleep. I'd love to believe the opposite, but I don't know of any evidence. But one thing—

Charlie Rose: Faith, Carl, faith.
Carl Sagan: One thing that it has done is to enhance my sense of appreciation for the beauty of life and of the universe and the sheer joy of being alive.

Charlie Rose: You had a healthy portion of that before this, but even you it happened to.
Carl Sagan: Oh, there's no question.

Charlie Rose: An appreciation—
Carl Sagan: Every moment, every inanimate object, to say nothing of the exquisite complexity of living beings. Yeah, you imagine missing it all and suddenly it's so much more precious.

Charlie Rose: May you live a long time. Thank you very much.
Carl Sagan: Thank you. It's a pleasure.

The Final Frontier?
Joel Achenbach / 1996

From the *Washington Post*, May 30, 1996. © 1996 The Washington Post. Reprinted with permission.

The man who answers the door does not look, at first glance, like Carl Sagan. The chemotherapy has eliminated the thick brush of black hair. He is bald, bony. He appears old, too old to be Carl Sagan.

But then the words rumble forth at a familiar frequency, low and deep, the syntax and vocabulary chosen with scientific precision. He takes a seat on the couch, and within minutes is speaking of the dimensions of the universe. Then comes a word, a verbal signature: billions.

The first consonant is explosive, a rocket whose payload is the soft vowel that follows.

Biiillyuns.

It's Sagan, all right.

Myelodysplasia, a life-threatening blood disease that causes a catastrophic failure of the immune system, has twice brought the celebrity scientist to the edge of death. He needed a bone marrow transplant and two rounds of chemotherapy. But he seems to have it beat.

"No myelodysplasia. No anomalous cells. Nothing," he says.

Sagan, ever the scientist, talks about his body in dispassionate, clinical terms. Of myelodysplasia he says, "There is some faint evidence that it is due to benzene and other aromatic hydrocarbons, but that's merely faint."

The man who wrote the "Life" entry in the *Encyclopaedia Britannica* and who has spent much of his career searching for life on other worlds has struggled with the very mundane, terrestrial problem of keeping his own heart beating.

It seems almost impertinent of Nature to confront Sagan with the question of life on such an individual scale. He has scrutinized images of Venus, Mars, the moons of Jupiter and Saturn. He has flown, empathetically, with robotic probes that have reconnoitered the outer limits of the solar system. He has pointed giant antennas at distant stars and tried to tune in radio signals from advanced galactic civilizations. He believes, to the extent that a strictly rational scientist holds beliefs at all, that life is abundant in the cosmos. He estimates that our galaxy alone holds one million technological civilizations.

And yet he now has to face the frustrating possibility that he will never be able to prove it. It seems likely to most scientists that, among the billions and billions of stars in each of the billions and billions of galaxies, there is life, even intelligent, technological, gregarious life that could transmit messages throughout space. But so far there's no trace, not even a microbe. Mars is a frozen desert. Venus is Hell. Our solar system is a collection of dazzling but inanimate objects, apparently lifeless but for the blue planet whose distance from the sun is in the narrow range where water neither instantly vaporizes nor turns to ice.

Sagan did, once, pick up some static with his alien-finding antennas. The provocative chirps sent chills up his spine, but the signals were not repeated. Was that an alien empire communicating? Or some meaningless experimental glitch? Sagan doesn't know.

"If we find it, it will revolutionize our knowledge of the universe and ourselves. If we don't find it after a really systematic search, then it underscores something about the rarity and preciousness of life," he says.

He'd like to find it.

"I'd rather there be extraterrestrial life discovered in my lifetime than not," he says. "I'd hate to die and never know."

As a boy, Carl Sagan would go into a field and lay his head on a log or a pillow and stare into space. He'd try to situate himself so he could see only stars, no trees or buildings, just the raw spectacle of the heavens.

Supine, he traveled in space. He is one of those people who do not view things like stars as fixed objects on a dome above us, twinkling cutely. He can feel the immensity of the universe, the raw power of stellar fusion, the violence of supernovas, the irreversible darkness of black holes. His gift is the ability to communicate a sense that all these planets and stars and galaxies actually mean something, that there is significance in ancient light from things so very far away.

Sagan often points out that every single one of the heavy atoms in our bodies—all of our carbon and oxygen atoms, for example—were once jettisoned from the interior of exploding stars. We are "starstuff," to use a classic Sagan word. This is not just a glib remark: It's Sagan's deeply felt connection to other worlds. Sagan can locate himself and his species in this immense place called the universe; it's his home.

No one of course can really envision it all, not even Sagan, who can only imagine little models of galaxies—"toys," as he puts it. Even the Sagan brain can't really picture billions and billions of stars.

"I can imagine that the Milky Way galaxy is over here and that the galaxy in Andromeda is over here and"—he is gesturing, making an invisible model in the air—"that they are a few centimeters away from each other. Here they are sitting in the air in front of me and then I can imagine the Magellanic clouds, which are satellites of our Milky Way. I know it contains four hundred billion stars, or whatever the right number is, but I surely don't have a picture of those four hundred billion in my head."

Some cultures, he points out, have no numbers bigger than three.

Sagan is probably the country's premier science popularizer (a term that is a pejorative among a certain furrowed-brow breed of scientist). When he became an astronomer he chose to study planets, even though at that time planetary science was considered a fringe field, damaged by the peculiar imaginings of astronomer Percival Lowell, who thought there were canals on Mars, the handiwork of Martian engineers. Serious astronomers studied distant galaxies, quasars, the background radiation that permeates the universe, the large-scale structure of the cosmos. "There was a kind of view that the seriousness of astronomy was proportional to the distance of the object," Sagan says. "The planets are too close."

Sagan made his mark early. In the 1950s his research helped show that Venus, under its thick cloud cover, is scorching. He probably peaked as an icon in 1980 when he hosted the epic PBS series *Cosmos*, but it is dangerous to try to summarize his career too quickly. This is a man who's won a Pulitzer Prize (for *The Dragons of Eden*, a book on the evolution of human intelligence), published a couple of hundred scientific papers, founded the Planetary Society for people interested in space science, written articles regularly for *Parade* magazine, and recently finished collaborating with his wife, Ann Druyan, on the screenplay for a movie based on his novel *Contact* (Jodie Foster will star). Oh yes, he also has an asteroid named after him, has won a Grammy for an audiocassette reading of his book *Pale Blue Dot*, has just put out a new book, *The Demon-Haunted World*, a polemic against pseudoscience, has finished a book of essays and is working with his wife on another novel, a romance. Plus there's his full-time job: professor at Cornell University in Ithaca, N.Y. The list goes on, a crushing output. It is hard to get a printed copy of his curriculum vitae, because it runs about two hundred and fifty pages. His office is happy to provide it in the form of two computer disks.

All that work, and he's most famous for three words: "Billions and billions..."

"I never said it at all," he says. "I never said 'billions and billions.' When we updated and reconfigured *Cosmos*, I had to go through the whole business and one of the things that I was watching is did I ever say it. And I never did."

Not only that, but he wouldn't say such a thing.

"It's so imprecise. How many is billions and billions? One or two? A hundred?"

While Sagan's interests have led him in many different directions, his abiding passion is the search for intelligent life in the cosmos. When he started as an astronomer in the '50s, he says, "an interest in life elsewhere was a disreputable idea." In the mid-1970s Sagan was one of the most vocal proponents of the notion that life might be detected on Mars.

Growing up, Sagan had read the John Carter books of Edgar Rice Burroughs. In the novels, Carter is suddenly, mysteriously, paranormally transported to the red planet, where he fights and romances amid a dying civilization. The science of Burroughs was thin, but Sagan never forgot Mars. He was part of the imaging team for the Viking lander in 1976; he wanted to study the pictures in case life was so precocious, so rambunctious, all you had to do to detect it was to look at it. He wanted Viking to have a flashlight in case Martian critters came out at night.

Colleagues thought he was a dreamer and laughed at such Saganesque notions as putting edible paint on the landers for the Martian life to lick. In the 1960s Sagan had gladly collaborated in the work of a Russian colleague who thought Phobos and Deimos, the quirky, potato-like Martian moons, showed signs of being artificial satellites, possibly the remnants of a Martian civilization gone extinct. Sagan's brand of science is full of possibilities, things not yet ruled out, marvels still conceivable. A 1976 *New Yorker* profile, not entirely flattering, quoted him as saying, "Someone has to propose ideas at the boundaries of the plausible, in order to so annoy the experimentalists or observationalists that they'll be motivated to disprove the idea."

The Viking lander found a sterile planet.

Was the problem simply that Viking landed in the desert? Could there be some form of life under the soil? Sagan's voice gets enthusiastic when he talks about the evidence that millions of years ago there were rivers flowing on Mars. Where there was abundant water there may have been abundant life. Dead life is better than no life at all.

Sagan has the distinction of co-authoring the first message to extraterrestrial beings. It was a gold-anodized plate affixed to the *Pioneer 10* spacecraft, launched in 1972 and bound for the asteroid belt, Jupiter and then interstellar space. The plate showed, among other things, a spacecraft emerging from the third of nine planets around a star. Sagan's wife at the time, Linda Salzman Sagan, added a line drawing of a nude man and woman; much public debate was aroused by the fact that the man had genitalia but not the woman.

It was while working on another message to aliens, the "Voyager record" placed on the two probes a few years later, that Sagan fell in love with Ann Druyan. She was creative director of the project, he the producer. They declared their love for one another on June 1, 1977.

"The revelation of being in love with each other was like the discovery of a scientific truth," Druyan says. "It was like Eureka, it was like Archimedes. It was like truth."

Sagan remembers the first time he listened to the Andromeda galaxy. It was in 1975. SETI—the Search for Extraterrestrial Intelligence—was a young and brash experiment.

Astronomers refer to the Andromeda galaxy as M31—the 31st in a series of nebulae catalogued by Charles Messier in the nineteenth century. For hundreds of years these nebulae were mere smudges in telescopes, their composition, dimension and significance yet unfathomed. Only in the third decade of this century did Edwin Hubble, the astronomer for whom the famed space telescope is named, discover that most of these smudges were choked with stars. They were galaxies, island universes far outside the confines of the Milky Way, a Holy Cow revelation of cosmic proportions.

So Sagan and a colleague, Frank Drake, aimed a radio telescope at M31 and listened on a particularly quiet frequency that would seem, for any intelligent species understanding the electromagnetic spectrum, an obvious choice for sending a how-de-do.

They heard only static.

"Okay, it's very far away, as we were saying before, so you have to have a very fancy civilization. But in one hundred billion stars there's not one civilization? I can't imagine. I can remember being not so much disappointed as surprised," Sagan says. "I thought, you know, I thought there ought to be, and there weren't."

He once told an interviewer that he was literally depressed for a week by the result.

More tantalizing were the results of Project Meta, a broader SETI search conducted in the late 1980s by Sagan and astronomer Paul Horowitz. On several dozen occasions they detected strong, brief electronic signals of . . . something. Most could be explained away as malfunctions of their instruments or interference from some terrestrial object, such as an airplane. But the five strongest signals came from the general direction of the center of the Milky Way galaxy.

He tells a reporter that the chance of this being accidental is "something like half a percent," and then hastens to add: "That's not strong enough to be sure. It's certainly suggestive. You know, it sends a kind of chill down your spine, your palms get moist, your breathing gets heavy."

Sagan has several possible explanations for why alien signals have proved so elusive. Maybe it's just the energy requirement of sending signals in all directions across such vast distances. Or maybe the aliens don't want to communicate with primitive creatures like us, and are intentionally bypassing the obvious frequencies in the electromagnetic spectrum, choosing instead a medium that we have yet to discover, like "Zeta waves." Sagan says, "I don't know what Zeta waves are, but they're much better than radio."

Or maybe: "No civilization survives long enough to develop power levels adequate to make such communications. All civilizations destroy themselves shortly after achieving a technological level consonant with radio astronomy."

Three years ago NASA canceled its SETI program. The search for extraterrestrial signals is entirely a private obsession now, largely funded by millionaires with an interest in making contact. But the optimists may have to deal with the possibility that this universe is not amenable to interstellar socializing. The fate of any intelligent species may be loneliness.

In the fall of 1994 Sagan was busy conducting "his usual five careers at a time," in the words of his wife, when she noticed a bruise on his arm that was slow to go away. She encouraged him to get a blood test. The doctor called Druyan while her husband was on the road.

"Is Carl in bed?" the doctor asked.

No, he's traveling, she said.

"That's a relief," the doctor said, "because these blood tests are the result of a gravely ill person. The person with these blood results couldn't possibly be on the road."

Sagan was retested. In December 1994, Sagan and Druyan were on a conference call with some Hollywood people, talking about the screenplay they'd written for *Contact*. They heard a beep signaling another call. They were expecting to hear from Sagan's doctor.

"I've got bad news for you," he said.

Myelodysplasia. Sagan had never heard of it. But the facts were clear: both his white and red blood cells were severely depleted, and he'd die if untreated. Might die anyway. He'd need a bone marrow transfusion.

They hung up with the doctor, got back on the line with the Hollywood people, and, Druyan says, continued talking about the movie.

Leaving his home in Ithaca, Sagan temporarily settled in Seattle to be treated at the Fred Hutchinson Cancer Research Center. His sister, Cari, donated the bone marrow he needed to stay alive. To prevent his body from rejecting the marrow, he had to take, in one sitting, seventy two pills labeled "BIOHAZARD." These essentially wiped out his immune system and would have killed him outright had he not had the bone marrow transplant immediately.

In the meantime he could have been killed by a single rogue microbe—some humble expression of the diversity of life.

Sagan seemed to have recovered from the disease when he learned, this past December, that he had fast-growing "anomalous" cells in his blood. Cancer, in other words. That meant more chemotherapy. He returned to Seattle. From his hospital bed he wrote a moving piece for *Parade* magazine: "There are scientific problems whose outcomes I long to witness—such as the exploration of many of the worlds in our solar system and the search for life elsewhere."

Sagan will not give up the dream of going to the stars. Maybe we can turn asteroids into spaceships, and mine them for energy sources as we trek across the void. Maybe our destiny is to evolve, among the stars, into something beyond human, becoming enlightened beings, a cosmic consciousness, the mind of the universe. Sagan is a visionary. But he also knows the cold, hard facts. For the foreseeable future, human beings are stuck on a rocky planet around a yellow sun on the Sagittarius spiral arm of the Milky Way galaxy. The credible goals for the human race are more limited than they were three decades ago.

Just staying alive, for starters.

In 1983 Sagan co-authored a highly publicized scientific paper arguing that nuclear war would culminate in a "nuclear winter" in which global temperatures would fall so dramatically that human life might become extinct. The paper inspired angry debate. Some accused Sagan of overstating the case. Eventually, more sophisticated computer models showed a less severe drop in global temperatures. Sagan and his colleagues had to revise their conclusion. Nuclear winter looked more like nuclear autumn. Sagan got the essence of the situation correct, but the error of proportion added to suspicions that anyone so smooth on TV must be a lightweight. (Sagan had been on Johnny Carson so many times, he became known as the Joyce Brothers of astronomy.)

In 1992, Sagan's name was one of sixty nominated for membership in the National Academy of Sciences. The other fifty nine made it without a hitch. But someone objected to Sagan.

Sagan's case was argued by Stanley Miller, a chemist who did pioneering work on the origin of life. He believes Sagan's scientific work, such as his research on the atmosphere of Venus, is often overlooked. The anti-Sagan faction countered that if the fluffy stuff of Sagan's career were swept away, there wouldn't be enough hard science underneath.

One member who was present says, "If he had not done television, he probably would be in the academy."

Sagan was voted down.

Sagan swears he doesn't dwell on the insult. He says he had assumed years earlier that he'd never get in.

"It seemed quite late," he says. "To discover that it was still a live issue surprised me more than learning that there were people opposed to my membership."

Druyan says of that period: "It was painful. It seemed like a kind of unsolicited slight. We hadn't done anything; he hadn't done anything. It was clear from people who were present at the time that there was something venomous about it."

It's just jealousy, she says. "I think there are few people who thought, 'I wrote a book; why wasn't that a bestseller?' "

Sagan concedes that his phenomenal range can be seen as a weakness. "It's a question of the balance between breadth and depth," he says. "Everyone has limitations of time and ability. Certainly it's true that if you spend a lot of time on breadth, you must be losing something on depth. . . . But I also see scientists who are bummed out after a while, and their productivity declines in their narrowly circumscribed field."

The academy's rejection of Sagan can be read as, if nothing else, a startling case of ingratitude. Sagan, more than almost any scientist alive, has tried to promote science, portray it as romantic and interesting, make people like it.

"There's a suicidal aspect of it," he says. "Here's science dependent as never before on public funds, and so continuing science depends on public support. And how's the public going to support it if they don't understand it?"

A couple of years later, the academy did make Sagan an honorary (and nonvoting) member when it gave him the Public Welfare Medal, an award for his educational efforts. The citation read: "No one has ever succeeded in conveying the wonder, excitement and joy of science as widely as Carl Sagan and few as well."

His new book, *The Demon-Haunted World: Science as a Candle in the Dark*, is his most coherent promotion of the scientific method. It's a collection of mini-essays on the boneheaded notions of pseudo-science, ranging from alien abductions to "recovered" memories of Satanic ritual child abuse. It's also darker and graver than Sagan's other work, with a touch of frustration, as though Sagan is astonished that despite two thousand and five hundred years of scientific inquiry since Aristotle there remain people who don't get it, who reject science in favor of myth, superstition, the paranormal.

The man who searches for their signals from outer space is alarmed that so many people think the aliens have beamed them aboard their flying saucers, performed surgical experiments on them or mated with them. Hasn't happened, Sagan says.

"A lot of the most fundamental physics can be written in the terms of prohibitive acts," Sagan says. "Thou shalt not travel faster than light. Thou shalt not measure the position and momentum of an electron simultaneously to whatever accuracy you want. Thou shalt not build a perpetual motion machine. . . . A lot of people—new agers, for example—are annoyed. They think everything can be done."

Sagan doesn't think everything can be done, but he does think everything can be questioned—even God. This is a delicate area for Sagan, who denies that he is an atheist.

"An atheist has to know a lot more than I know. An atheist is someone who knows there is no God."

When he wrote of his disease in *Parade* he received hundreds of letters, many of them challenging him for questioning the existence of a Creator and life after death. They told him that someday he will die and will find himself before God. They asked: "What are you going to say to Him?"

Sagan already knows: "What took you so long?"

For Sagan, it's simple: a scientist needs evidence of things. Faith is not part of the game.

The axiom applies in matters both great and small. One day while Sagan was talking long-distance to a reporter there was, in the background, the sound of a doorbell. An exterminator had stopped by to spray for carpenter ants. Sagan could be overheard grilling the guy:

"What are you spraying? What chemical? You know its structure? You know its chemical formula?"

The exterminator gave the name of a chemical.

"That's just a name," Sagan says. "You have a structural diagram of the molecule?"

The exterminator eventually produced a diagram. Sagan approved the molecule.

In many ways, Sagan is already a man of the past. As he looks forward in time, and out into space, he is one of the guiding spirits of the Space Age—which, in a sense, is already over, a historic period starting in the late 1950s and ending sometime in the 1970s or 1980s. In 1962, NASA had a plan to send an eight-man mission to Mars at the end of the 1970s. In his 1973 book *The Cosmic Connection*, Sagan predicted that there would be semi-permanent moon colonies by the 1980s, with moon children eventually referring to Earth as "the old country." His optimism was nothing compared with Stanley Kubrick's: In the late 1960s, according to Sagan, the filmmaker asked Lloyds of London to insure *2001: A Space Odyseey* against the possibility that extraterrestrial life would be discovered during the filming of the movie. Kubrick feared that alien contact would ruin the film's plot, in which contact is made in 2001.

Now such dreaming seems so quaint. NASA put its money into the go-nowhere space shuttle. The Space Age is a '60s conceit, and the term itself is kind of campy. (We've moved on to the Information Age.)

Sagan has written that human beings are "like a toddler who takes a few tentative steps outward and then, breathless, retreats to the safety of his mother's skirts."

Yet it was not just lack of courage that halted manned exploration of outer space; there were fiscal, political and even astrophysical realities that caused our retreat. The Apollo program, everyone now realizes, was an extension of the Cold War, and in the post-Soviet era there is no short-term political or economic reason to spend $100 billion to go to Mars or any other distant world. Space has apparently become inaccessible again. The Apollo astronauts are old and gray.

Still, Sagan refuses to be disappointed by the unrealized expectations of the Space Age.

To explain why, he ticks off what he believes are the three major scientific revelations in planetary science in the post-Apollo era. Only the first, the dearth of obvious life forms in our own solar system, is disheartening. But the second discovery is that space is permeated with organic molecules—that is, carbon molecules, big, heavy structures thought to be essential or at least highly conducive to the origin of life.

"Comets are made one-quarter of organic matter. Many worlds in the outer solar system are coated with dark organic matter. On Titan, organic matter is falling from the skies like manna from Heaven. The cold, diffuse interstellar gas is loaded with organic matter," he says. "There doesn't seem to be an impediment about the stuff of life."

And then comes the third revelation: That this stuff of life has plenty of places to land, accrete, do its business of turning into self-replicating, mutating, evolving organisms. Astronomers now have abundant evidence that planetary systems are commonplace around stars. Five new jumbo planets— or objects that behave very much like planets—have been discovered around other stars in just the past six months.

So Sagan is bullish on life—as optimistic as he was in the 1960s.

"Nothing has changed," he says.

Sagan holds out the hope that there could be life on the Jovian moon Europa, or on Saturn's huge moon Titan. Or maybe the probes now voyaging to Mars—there are about twenty unmanned missions from various countries planned in the next few years, a veritable armada of spaceships—might unearth (so to speak) the signs of ancient Martian life.

And if not, Sagan remains philosophical. He says the current absence of evidence of life elsewhere in the solar system is actually an additional motivation to send humans to other planets; the sterility of those environments eliminates the danger that we might inadvertently kill precious, exotic alien life forms by infecting them with stowaway microbes. And he's been trumpeting the practical benefit of settling on other worlds as an insurance policy against a possible catastrophic impact on the Earth by an asteroid or comet.

"As nearly as we can tell, so far at least, there is no other life in this system, not one microbe. There's only earthlife," he writes in *Pale Blue Dot*. What follows is a classic Saganism: "In that case, on behalf of earthlife, I urge that, with full knowledge of our limitations, we vastly increase our knowledge of the solar system and then begin to settle other worlds."

Someone has to speak for earthlife. Might as well be Carl Sagan.

A few weeks ago Sagan returned home to Ithaca, his blood scoured of anomalies, his hair growing back. He knows he could get sick again. He could die. Druyan says she's had the wits scared out of her. She says she's betting her husband will make a full recovery, because he has so much to live for—so many unanswered questions.

He is working feverishly again. He has worked on a paper titled "On the Rarity of Long-Lived, Non-Spacefaring Galactic Civilizations." His laboratory is trying to re-create the atmosphere of Titan. There may not be life but there are lots of organic molecules. "Titan's tremendously exciting in that context," says Sagan.

In the meantime he remains grateful for earthlife: his own, his family's. After sixty-two fascinating years, he has five children, including a five-year-old son and thirteen-year-old daughter, and a wife of whom he once wrote: "In the vastness of space and the immensity of time, it is still my joy to share a planet and an epoch with Annie."

On this planet, in this epoch, Carl Sagan's search for life goes on.

Index

ABC News, xxiii
Acquired Immune Deficiency Syndrome (AIDS), 100, 129
Advanced X-Ray Astronomical Facility (AXAF), 105
Agent Orange, 101
agnosticism. *See* God
Alexander, Lynn. *See* Margolis, Lynn
alien abduction, xi, xvii, 99, 101, 103–4, 131, 146, 159. *See also* UFOs
Alpha Centauri, 6
alternative energy, 92
Ambrose, Stephen, 143
American Association for the Advancement of Science, xxiii
American Astronomical Society, xxiii
American Institute of Physics, xviii–xix
Ames Research Center, 25
Anderson, Don L., 21
Andromeda galaxy, 155
Apollo Achievement Award, xxii
Apollo program, 55–56, 83–86, 109, 161
Aquinas, Thomas, 18
Archimedes, 155
Arecibo Observatory, 8, 35
Aristotle, 12, 159
artificial intelligence, xiv
Asimov, Isaac, 8
asteroid, comet, and meteor collisions, xvi, 82, 85–87, 93–94, 105, 108
astrology, xvii, 62, 99, 101, 127, 141–42, 145
Astrophysical Journal, 90
atheism. *See* God
Atmospheres of Mars and Venus, xxi
Augustine of Hippo, 64
Aztec, 16

Bach, Johann Sebastian, xii
baloney detection kit, xvii

Barbarella, 48
Barnard's star, 5
Bay of Pigs, 55–56
Berendzen, Richard, xxiii
Bermuda Triangle, 47, 54–55
Berry, Chuck, 49
Billions and Billions, xviii, xxv
Biology and the Exploration of Mars, 33
Blob, The, 48
Bradbury, Ray, xxiii
Brand, Stewart, 24
Brave New World, 123
British Interplanetary Society, 28
Broca's Brain, xxiv, 57
Brothers, Joyce, 158
Bryan, Robert, 91
Buddhism, 65–66
Budiansky, Stephen, ix–x, xix–xx
Burroughs, Edgar Rice, x, 27–28, 50, 154
Bury, J. B., 10
Bush, George H. W., 79, 85

California Institute of Technology, xxiv, 96, 141
Carl Sagan: A Life in the Cosmos, xii, xx
Carl Sagan Productions, 48
Carnegie Mellon University, xxiv
Carson, Johnny, xii, xviii, 25, 49, 158
Caruso, Enrico, 17
Cassini program, 96
Chariots of the Gods?, xxiii
China Syndrome, The, 138
Civil War, The, 106
Clarke, Arthur C., xxiii, 16, 29
Clinton, Bill, 102–3, 142
Close Encounters of the Third Kind, 47–50. *See also* alien abduction; UFOs
Cold and the Dark, The, xxiv
cold fusion, 133

Index

Cold War, 7, 9, 42, 83–84, 103, 109. *See also* nuclear disarmament; Union of Soviet Socialist Republics (USSR)
Collins, Glenn, xx
Comet, xxiv, 100
comet collisions. *See* asteroid, comet, and meteor collisions
Committee for the Scientific Investigation of Claims of the Paranormal (CSICOP), 99
Communication with Extraterrestrial Intelligence, xxiii
Communication with Extra-Terrestrial Intelligence (CETI), 34. *See also* Search for Extra-Terrestrial Intelligence (SETI)
Condon Lectures, xxii
Contact (film), xvii–xviii, xxv, 89–90, 123–24, 139–40, 153
Contact: A Novel, xvii–xviii, xxiv–xxv, 89–90, 94, 97, 123–24, 139, 153
Coppola, Francis Ford, 24
Cornell University, xxii, 3, 20, 28, 33–35, 48, 57, 68–69, 82, 99, 125, 140–41, 153–54
Cosmic Connection, The, xii, xviii, xx, xxiii, 24, 27, 34, 36, 48, 51, 57–58, 68, 160
Cosmos (book), xiii, xxiv, 57, 99
Cosmos (miniseries), xii–xiii, xxiv, 48, 54–55, 57–68, 99, 106, 125, 138–39, 141, 153–54
creationism, 60–61, 102, 142
crystal healing, xvii

Däniken, Erich von, xxiii
dark matter, 98
Darth Vader, 48
Darwin, Charles, 61, 104
David Duncan Chair in Astronomy and Space Sciences, xxiii, 33, 57, 82, 99, 125, 141
Dawson, Jim, xiv, xx
De Kamp, Peter Van, 6
Deimos, 154
democracy, xv, 63–64, 72, 100, 105, 128, 130, 135, 137
Demon-Haunted World, The, xvii, xxv, 113–14, 125–26, 132, 140–41, 146, 153, 159
Discovery of Our Galaxy, The, 12
Dragons of Eden, The, xxiii, xxv, 51, 57, 63, 68, 125, 153
Drake, Frank, xxiii, 33–35, 155
Dreyfuss, Richard, 47
drugs, 14–15, 62, 100

Druyan, Ann (wife and collaborator), xviii, xxiii–xxiv, 58, 62–64, 67, 89, 100, 153, 155–58, 162
Dylan, Bob, 64
Dyson, Freeman J., 39

education. *See* science education
Ehrlich, Paul R., xxiv, 136
Einstein, Albert, 14, 17, 40, 69–70, 119, 132–33
Eliot, T. S., 58, 67
Emmy award, 99
Encyclopaedia Britannica, 151
Enlightenment, the, xvii
environmentalism, xiv–xvi, xix, xxi, xxiv, 60, 77–81, 86, 88, 92, 97, 100–3, 105, 107–10, 117, 123, 128
E. T.: The Extra Terrestrial, 138
Europa, 161
Evolution (journal), xxi
extraterrestrial life, x–xii, xvii–xix, 3–12, 15–20, 22–27, 29–35, 39–40, 47–53, 91, 96–97, 104, 112, 124, 131, 151–52, 157, 161–62

Fermi, Enrico, 29
Ferris, Timothy, xii, xxiii
Feynman, Richard, 43
51 Pegasus, 118
Flatow, Ira, xvii
Foster, Jodie, xvii, xxv, 89, 123, 153
Fresh Air. *See* NPR *Fresh Air*
Freud, Sigmund, 59, 120–21

Gagarin, Yuri, 55–56
Galileo program, 99, 109
General Agreement on Tariffs and Trade (GATT), 83
ghosts, 145
Gibbons, Jon, 142
Gierasch, Peter, 34
Gingrich, Newt, 142
Global Forum of Spiritual and Parliamentary Leaders on Human Survival, xxiv
God, xiv, 61, 68–71, 75, 132, 159–60
Gore, Al, 102–3, 142
Greek model of the universe, 59
Greene, Cari Sagan (sister), xvii, 116, 149, 157
Gross, Terry, xiv, xx

Haldane, J. B. S., 29

Index 165

Halley's comet, 44
Han Solo, 50
Harvard University, xxii, 28, 33–34, 90
Hitler, Adolf, 114, 137
Hollywood, 47–50
Horowitz, Paul, 90–91, 156
Hubble, Edwin, 155
Hubble Space Telescope, 104–5
Huxley, Aldous, 123
Huygens, Christiaan, 27

Icarus (journal), xxii, 24
Idea of Progress, The, 10
Indiana University, 29
Intelligent Life in the Universe, xxii, 3, 6, 24
International Astronomical Union, 30
International Council of Scientific Unions, 42
International Space Station, 87
Io, 59

Jefferson, Thomas, 143
Johnson, Lyndon B., 4
Jupiter, xvi, 5, 7–8, 11, 18, 37, 51, 53, 59, 82, 86–87, 106, 118, 151, 155, 161

Kalosh, Anne, xi, xv, xx
Kant, Immanuel, 104–5
Kellogg, W. W., xxi
Kennedy, Donald, xxiv
Kennedy, John F., 55–56, 85
Knight, J. Z., 148
Kondratyev, Kirill, xxii
Kubrick, Stanley, 50
Kuiper, Gerard, 31

Lederberg, Joshua, 22–23, 32–34
Leonard, Jonathan, xxii
Lewis, Meriwether, 143
ley lines, 145
Life Beyond Earth and the Mind of Man, xxiii
Loch Ness monster, 148
Lomberg, Jon, xxiii
Lowell, Percival, 12, 30–31, 153
Lucas, George, 50

Mack, John, 146
Maclaine, Shirley, 148
Man and the Cosmos. See *Cosmos*
Margolis, Lynn (Sagan's first wife), xxi, 31

Mariner program, xviii, 3–4, 12, 33, 48, 57, 99, 109
Mars, x, xii, 3–4, 12, 19–35, 42, 44, 52, 78, 81, 84–85, 87–89, 93, 95, 104, 110–11, 125, 130, 151–54, 160–61
Mars and the Mind of Man, xxiii
Massachusetts Institute of Technology, 23
McDonald Observatory, 31
Mercury, 109, 118
Meredith, Dennis, xiii
Mesmer, Franz, 114
Messier, Charles, 155
meteor collisions. *See* asteroid, comet, and meteor collisions
Miller, George, 89–90
Miller, Stanley, 30, 32, 158
Miller Research Fellowship, xxi
Minneapolis Star-Tribune (interview), xiv, xx
Montagu, Ashley, xxiii
Morrison, Philip, xxiii, 23, 26
Mithra, 11–12
Muller, H. J., 29
Murmurs of Earth, xxiii, 68
Murray, Bruce, xxiii
music, 49
Mutch, Tim, 21–22
myelodysplasia, xvi–xviii, xxv, 116, 126, 148–50, 151, 156–58, 162
mysticism, 14–15, 62

National Academy of Sciences, xxi, 33, 92, 125, 158
National Enquirer, The, 49
National Science Foundation, 35, 141
Natural History of the Mind, The, 62
near-death experiences, 71–72, 149
Neptune, 8, 82, 109
New York Times, 4, 66
New York World's Fair (1939), x
New Yorker, 154
Newton, Isaac, 11, 133
Nixon, Richard, 42
nuclear disarmament, xiv–xvi, xxiv, 72, 99–100, 114, 118, 158

Obst, Linda, xiv, xix–xx, 90
Oparin, Alexander, 29, 32
Oppenheimer, J. Robert, 72
Oregon State University, xxii

Origin of Life on Earth, The, 32
Orion Nebula, 104–5
Other Worlds, xxiii, 24
overpopulation, 116–17, 123
Owen, Tobias, xxii

Page, Thornton, xxii
Pale Blue Dot, xv–xvii, xx, xxv, 82–83, 92, 100, 106–7, 153, 162
Parade magazine, 153, 157
Path Where No Man Thought, A, xxiv
Patriot missiles, 136
Pauling, Linus, 136
Peabody Award, xxiv, 99
Persian Gulf War, 80, 136
Phobos, 154
Physical Studies of the Planets, xi, xxi
Pierce, Ponchitta, xiv–xv
ping pong, 42
Pioneer program, 18, 33, 44, 51, 155
Planetary Atmospheres, xxii
Planetary Exploration, xxii
Planetary Society, xxiv, 82, 90, 99, 141
Planets, The, xxii, 5
Pluto, 49
Pollack, James B., 25, 34
Postal, Ted, 136
Poundstone, William, xii, xx
Princeton University, 28
Prix Galabert, 34
Project BETA, 91
Project Blue Book, xxii
Project META, 90–91, 156
psychology, 120–21
Public Welfare Medal, 159
Pulitzer Prize, xxiii, 57, 68, 99, 125, 141, 153
Pythagoreans, 129

Reagan, Nancy, 101
Reagan, Ronald, 99, 102
religion, x, xiii–xiv, xix, 4–5, 14–15, 58–59, 61, 66–75, 102, 120, 122–23, 126–27, 132–33, 135, 144–45, 147–50. *See also* God; Sagan, Carl, core beliefs
Rensberger, Boyce, xi, xx
Rightmyer, Jack, xix
Roberts, Walter Orr, xxiv
robotics, xxiii, xxiv

Roswell incident, 130–31
Rycroft, Michael, xxii

Sagan, Alexandra (Sasha) (daughter), xxiv
Sagan, Cari. *See* Greene, Cari Sagan
Sagan, Carl: childhood, ix–xi, 26–28, 50, 68, 99, 119–20, 127, 152, 154; college years, 28–33, 99; conversations with children, 20–22, 94–95; core beliefs, ix–x, xiii–xvii, xxiii, 61–62, 75, 82–83, 106–7, 122, 147–50; death (*see* myelodysplasia); philosophy of science, ix–xi, xiii–xvii, xxiii, 12–15, 36–38, 40, 47–48, 54, 57, 60–61, 64, 66–67, 74–75, 117–20, 123, 128–30, 132–35; public persona, xi–xix, xxiii, 19–20, 23–25, 48, 54, 117, 152–54, 158–59
Sagan, Dorion (son), xxi, 31
Sagan, Jeremy (son), xxi, 31
Sagan, Nicholas (son), xxii, 20, 25, 27
Sagan, Rachel Gruber (mother), x, xxi, xxiv, 147
Sagan, Samuel (father), xiii, xxi, xxiv, 147
Sagan, Samuel (son), xxiv
Sagan, Sasha. *See* Sagan, Alexandra (Sasha)
Salzman, Linda (second wife), xii, xxii, xxiv, 24, 33, 155
Saturn, 7–8, 33, 53, 82, 106, 151, 161
Schiaparelli, Giovanni, 30
science. *See* Sagan, Carl, philosophy of science
Science (journal), xv
science education, xv, 100–2, 105, 114–16, 119, 121–22, 127–28, 135–38, 140–44
science funding, xvi, 83–85, 93–94, 103–5, 110
Scientific American, 124, 138
Scientific Literacy Index, 100
Scopes trial, 17
scuba diving, 42
Search for Extra-Terrestrial Intelligence (SETI), xxiii, 48, 56, 90–91, 104, 124, 131, 152, 155–56. *See also* extraterrestrial life; Project META
Search for Life on Mars, The, 19–35
Shadows of Forgotten Ancestors, xxiv, 100
Shapley, Harlow, 28
Shklovskii, I. S., xxii, 3, 24
Shoemaker-Levy 9 (comet), xvi, 85–87. *See also* asteroid, comet, and meteor collisions
Sightings, 126
Simon, Pierre, 104–5
Simpson, O. J., 121
Sirius, 5–6

Index

skepticism. *See* Sagan, Carl, philosophy of science
Sky from Elsewhere, The (proposed children's book), 7
Smith, Harlan J., xxii
Smithsonian Institution, xxii–xxiii, 33
Soter, Steven, 58
Soviet Union. *See* Union of Soviet Socialist Republics (USSR)
Space Exploration Initiative, 85
Space Research XI, xxii
space shuttle program, xv, 43, 55–56, 84, 107–8, 160
"Spaceship of the Imagination," xiii, 55
Spielberg, Steven, 47–50
Spinoza, Baruch, 69, 132
Spitzer, Lyman, 28–29
Sputnik program, 115–16
Stalin, Joseph, 114
stamp collecting, 42
Stanford University, 22, 33
Star Wars, 49–50
Stendhal, Krister, xxiii
Strategic Defense Initiative (SDI), 99, 111
string theory, 134–35
Sullivan, Walter, xxiii
Sumerians, 15
Sunderland, Luther, 60
Sununu, John, 79
superconducting supercollider, 129
Swathmore Observatory, 6

Tau Ceti, 34
Taylor, Gordon Rattray, 62
terraforming, 34–35, 88–89, 107
Thing, The, 48
theory of relativity, 40–42, 44–46, 60
Thompson, William Irwin, 15
Three Mile Island, 138
Time, 34
time travel, 43
Titan, 35, 53, 95–96, 111, 161–62
Tonight Show, The. *See* Carson, Johnny
Toynbee, Arnold J., 16
Trumbull, Doug, 47
Tsiolkovsky, K. E., 59
20/20, xxiii
2001: A Space Odyssey, 50, 139–40, 160

UFOs, xi, xxii, 5, 15–16, 43, 47–49, 54–55, 99, 103–4, 113–14, 125, 130–31, 134, 145–46. *See also* extraterrestrial life
UFOs: A Scientific Debate, xxii
Union of Soviet Socialist Republics (USSR), 7–9, 19, 42, 99–100, 103, 114, 161
United Nations Earth Summit (1992), xiv–xv, 79
University of California, Berkeley, xxi–xxii, 33
University of Chicago, xi, xxi, 28–33, 99
University of Pennsylvania, 25
University of Texas at Austin, 134
University of Wisconsin, 30–31
Uranus, 8, 82, 106
Urey, Harold, 29–30
USSR. *See* Union of Soviet Socialist Republics (USSR)

Van Allen radiation belt, 37
Velikovsky, Immanuel, xxiii, 36–39
Venus, xiv, xxi, 34, 37, 42, 77–78, 81, 89, 109–10, 117–18, 151–52, 158
Veverka, Joseph, 34
Viking program, 19–26, 33–35, 44, 48, 52, 57, 88, 99, 104, 109, 111, 125, 154
Vishniac, Wolf, 33
Voyager program, xii, xv–xvi, 49, 51, 56–57, 77, 82, 96, 99, 106, 109, 111, 155

Wakin, Edward, x–xi, xiii–xiv
Wald, George, xxiii
Wallace, Alfred Russel, 29
War of the Worlds, 4
Weinberg, Steven, 134
Welles, Orson, 4
Whitehead, Alfred North, 63
Whitney, Charles, 12
Whole Earth Catalog, The, 24
Wildt, Rupert, 37
women in science, 138–39
World War II, 28
Worlds in Collision, xxiii, 38–39

Yale University, 33
Yerkes Observatory, 29

Ziegfeld Theater, 47
Zubrin, Martin, 89

PS 3569
.A287
Z475
2006